T0291733

# Quantum Strangeness

# Quantum Strangeness

## Wrestling with Bell's Theorem and the Ultimate Nature of Reality

George Greenstein

The MIT Press
Cambridge, Massachusetts
London, England

This book was set in Stone Serif by Westchester Publishing Services.

Library of Congress Cataloging-in-Publication Data

Names: Greenstein, George, 1940- author.
Title: Quantum strangeness : wrestling with Bell's Theorem and the ultimate nature of reality / George S. Greenstein ; foreword by David Kaiser.
Description: Cambridge, MA : The MIT Press, [2019] | Includes bibliographical references and index.
Identifiers: LCCN 2018043232 | ISBN 9780262039932 (hardcover : alk. paper)
ISBN 9780262549301 (paperback)
Subjects: LCSH: Bell, J. S. | Quantum theory. | Physics--Philosophy.
Classification: LCC QC174.12 .G7325 2019 | DDC 530.12--dc23 LC record available at https://lccn.loc.gov/2018043232

To Guy Blaylock and Arthur Zajonc
Close friends, valued colleagues, and fellow Wrestlers

"I think I can safely say that nobody understands quantum mechanics."
Richard Feynman, *The Character of Physical Law*

# Contents

Foreword by David Kaiser xi
Acknowledgments xvii

1. **The Great Predictor** 1

**Background to Bell**

2. **Silence** 11
3. **Half a Theory?** 17
4. **The Solvay Battles** 21
5. **Spin** 29
6. **An Impoverished Language?** 33
7. **The EPR Paradox** 37
8. **Hidden Variables** 41

**Bell's Theorem**

9. **A Hidden Variable Theory** 45
10. **Bell's Theorem** 55
11. **Stigma** 63

**Experimental Metaphysics**

12. **Experimental ...** 71
13. **... Metaphysics** 93
14. **Nonlocality** 97
15. **Quantum machines** 101
16. **A New Universe** 113

Appendix 1: The GHZ Theorem 121
Appendix 2: Further Reading 125
Notes 129
Index 135

## Foreword

Physics has more than its share of mind-bending ideas: the slowing clocks and shrinking meter sticks of relativity; enormous coagulations of matter, like black holes, that can rupture space-time itself. Yet the strangest ideas of all are clustered in quantum theory, physicists' remarkably successful description of matter and energy at atomic scales. Here we find descriptions of objects that seem to act as if they were in two places at once; of particles that can tunnel through walls; of Erwin Schrödinger's twice-fated cat, trapped in a zombie-like state of being both alive and dead. For all that, Schrödinger himself declared one idea in particular, quantum entanglement, to be "*the* characteristic trait of quantum mechanics, the one that enforces its entire departure from classical lines of thought."[1]

Schrödinger had done so much to contribute to quantum theory; his "wave function," obeying an equation he first published in 1926, remains central to physicists' efforts to describe quantum systems quantitatively. Almost a decade later, in 1935, Schrödinger coined the term "entanglement," though by then his enthusiasm for quantum theory had begun to wane. That same year his friend Albert Einstein teamed up with two younger colleagues, Boris Podolsky and Nathan Rosen, to issue his own, latest challenge to quantum theory. In their famous "EPR" paper (named for the authors' initials), they described a system involving a pair of entangled particles emitted from a central source. Physicists could perform various measurements on one particle, and thereby learn something about the second particle, far off in the distance. Indeed, the EPR authors concluded, physicists should be able to glean *more* information about the far-away particle than could be accounted for within quantum theory.[2]

Each particle, it seemed self-evident to Einstein and his coauthors, should possess definite properties on its own, independent of what physicists happened to choose to measure. If physicists elected to measure the first particle's position at a given moment, for example, they would learn about the position of the second particle, which had headed off at the same speed but in the opposite direction from the first particle. Or the physicists might choose to measure the momentum of the first particle, and thereby learn about the second particle's momentum. But surely the second particle had definite values for these and any other properties the physicists might have chosen to investigate, regardless of what choices the physicists had made. After all, Einstein's own theory of relativity made clear that no signal or influence could travel faster than the speed of light—so nothing the physicists might have chosen to do to the first particle should have been able to affect the second particle, which had traveled so far away. If relativity really set an absolute speed limit on how quickly *A* could influence *B*, then the second particle would need to carry all its own information with it, as it traveled through space; there would be no time to receive an update on what values for various properties it should have, based on the outcomes of measurements on the first particle. Therefore, the EPR authors concluded with a flourish, there existed "elements of reality"—real, definite properties of that second particle—about which quantum theory offered no description. Quantum mechanics, they argued, was incomplete.[3]

Within weeks a response came from Niels Bohr, the Danish physicist who had helped to craft quantum theory and who served as a kind of spokesperson for the emerging work. Bohr's response to EPR was rapid, but abstruse; to this day, it remains difficult to parse Bohr's argument. Central to his response, however, was a denial that objects in the subatomic realm really must carry complete sets of properties on their own. Rather, Bohr insisted, a particle might have *no* definite value of, say, momentum, until subjected to a particular measurement—as if a person had no definite weight until stepping on a bathroom scale. (Years later, Einstein famously asked a colleague if quantum physicists really believed that the moon was only there when someone chose to look.) Most important to physicists at the time, it seems, was that Bohr had responded at all. More recent scholarship has indicated that Einstein and Bohr were largely talking past each other—a series of miscommunications exacerbated by the rise of fascism in Europe, which had driven Einstein to emigrate to the United States, thereby ending the late-night, face-to-face discussions that the two had enjoyed during happier times. Each physicist died, decades later, having failed to convince the other when it came to quantum entanglement.[4]

The debate seemed to linger, with no clear resolution, for years. A young physicist from Northern Ireland, John S. Bell, shared many of the reservations about quantum theory that Einstein had expressed. Indeed, Bell had nursed a private concern about the topic since his student days, growing frustrated when students and teachers seemed to parrot Bohr's responses. Bell was quickly counseled to keep such "philosophical" concerns to himself—by the 1950s, quantum mechanics had moved to the very center of physicists' expanding efforts to understand everything from nuclear reactions to superconductivity to the properties of little devices like transistors. In every single case, the equations of quantum theory provided a remarkable match to experiments. So why, Bell's teachers pressed, should they continue to fret over the abstract questions that had distracted old-timers like Einstein and Bohr?[5]

Bell dove into mainstream topics for his research in high-energy physics, even as his thoughts kept returning to nagging questions about quantum entanglement. Finally, during a sabbatical in the United States in 1964, Bell brought many of his ideas to fruition. He tweaked the famous EPR thought experiment, focusing on specific combinations of measurements that physicists could perform on each of the two entangled particles. In just a few lines of algebra, Bell demonstrated that Einstein's pair of assumptions—that particles carry definite properties on their own, prior to and independent of measurement, and that no influence can travel faster than light—*necessarily* led to a contradiction with quantum theory. Bell identified a quantitative upper limit for how often the outcomes of certain combinations of measurements on the two particles could ever line up, if they behaved in accordance with Einstein's assumptions. If, instead, the particles were governed by quantum theory, then the measurements on each particle should be more strongly correlated, surpassing the upper bound that Bell had derived. If quantum theory were true, in other words, then performing a measurement over here really would seem to affect the behavior of some other tiny bit of matter, observed arbitrarily far away.[6]

On paper, Bell showed, the contrast was as clear as day: an Einstein-like world limited to one side of the bound; a quantum-mechanical world clearly surpassing that limit. The central question that had absorbed Einstein and

Bohr for decades could be posed in a laboratory, not just debated in a smoke-filled room. Bell published his paper, and then... nothing. Years went by before he heard so much as a peep of interest from the physics community.

In time, Bell's elegant paper happened to catch the eye of a few unconventional physicists, who recognized the magnitude of Bell's achievement. If they followed Bell's reasoning, and really conducted experiments of the sort he had described, they might be able to learn something deep about how the world works. Pioneering physicists like Abner Shimony, John Clauser, Michael Horne, Stuart Freedman, Alain Aspect, and a handful of others began to realize that by testing Bell's inequality in a laboratory, they could subject abstract, metaphysical mysteries to experimental investigation. Hence the term used in George Greenstein's lovely book: *experimental metaphysics*.[7]

For nearly half a century, physicists have subjected Bell's inequality to experimental test. Every single published result has been consistent with the predictions of quantum theory, showing correlations among measurements on pairs of entangled particles in excess of Bell's bound. Yet from the start, Bell, Shimony, Clauser, Horne, Aspect, and others have recognized that each of these tests has been subject to one or more "loopholes": little conceptual escape hatches by which an Einstein-like interpretation could still account for the experimental results.

Perhaps the particles, or other elements of the apparatus, had shared information (at or below the speed of light) during a particular series of measurements, thereby arranging for the measurements to line up the way they did. Or perhaps the detectors that had been used to measure the particles were inefficient, and failed to register any definite outcome some fraction of the time. Then it would be possible, at least in principle, for the strong correlations that showed up in those measurements that *were* successfully recorded to be but a rare statistical hiccup, some fluke that would have been washed out had all of the particles actually been measured. Or perhaps the striking correlations arose from some common cause, deep in the experiment's past, which had somehow nudged the selection of measurements to be performed and tipped off the particles in advance. After all, as Schrödinger himself acknowledged back in 1935, one should hardly be surprised when a student aces an examination, if she had received a copy of the questions ahead of time.[8]

Experimental groups around the world have tackled these loopholes, usually one at a time, since the early 1980s.[9] Only as recently as 2015 have various groups succeeded in measuring significant violations of Bell's inequality while closing not one but two of the stubborn loopholes.[10] More recently, my own colleagues and I have conducted experiments to address that strange, third loophole—the one regarding the seemingly random selection of measurements to perform on the entangled particles—by turning the universe itself into a pair of random-number generators. Yet again we have found—as have our colleagues around the world—the strong correlations that Bell had first identified.[11]

Why go through all the trouble? Aren't the dozens of previous experiments sufficient for us to declare the question settled? More than just stubbornness is at stake. In fact, quantum entanglement now lies at the center of a booming field dubbed "quantum information science," which promises major new devices. Quantum encryption and quantum computers—each of which has progressed well into the beta-testing stage—will function as promised only if entanglement is *real*. If, for some reason, entanglement were merely an artifact, and the world really were governed by Einstein's assumptions, then quantum encryption simply would not be secure, and quantum computers would fail to deliver the anticipated exponential speed-up compared to ordinary machines. These days, practical technologies built around quantum entanglement constitute a multibillion-dollar industry—contributing a new set of imperatives to keep testing Bell's inequality, in addition to the deep, metaphysical questions.[12]

Despite all the theoretical progress over the past half century, and the recent experimental advances in addressing various loopholes, we are still left with vexing questions. How could the world really *work* that way? How could two little specks of matter act in concert, even after they had moved arbitrarily far apart? In recent years, physicists have gone to extremes trying to articulate, in everyday scenarios, what entanglement implies about the world. There have been elaborate fables about cops chasing quantum robbers; quantum soufflés that do or do not rise; tales of twins who order drinks in bars an ocean apart; and so on.[13]

Amid all these discussions, George Greenstein's book is a special delight. With patience and clarity, Greenstein guides readers along this extraordinary conceptual journey. We are neither hushed nor rushed; never reprimanded, as John Bell himself had been as a student, to simply take this or that result on faith. Rather, Greenstein shares his own earnest struggles to come to grips with the ideas, to sit with them, to try to puzzle through what they might imply about the workings of nature at its most fundamental. And he does all this with simple illustrations and virtually no mathematics. His book is an invitation and a primer for those new to the topic, and a timely reminder to fellow specialists, who have long since grown comfortable with the mathematical formalism of quantum theory, that this central element of physicists' toolkit retains a beguiling strangeness at its core.

David Kaiser
Germeshausen Professor of the History of Science
and Professor of Physics
Massachusetts Institute of Technology
October 2018

## Acknowledgments

This book is the product of many minds.

Guy Blaylock, Danny Greenerger, David Harrison, David Kaiser, Preston Stahley, and Arthur Zajonc read the manuscript and gave me the immeasurable benefit of their advice. Andrew Fraknoi and John Harte read an earlier article and similarly gave me much help. And over the years I have benefited greatly from—and enjoyed—illuminating conversations with innumerable colleagues and friends: some of them scientists working in the field, others nonscientists who nevertheless had insights that gently prodded me in productive directions.

Renate Bertleman, John Clauser, and Jain Wei Pan kindly contributed their own personal photographs of people and equipment.

To all my deepest thanks.

# 1 The Great Predictor

It was many years ago that I first encountered the Great Predictor.

I was thrilled to meet him. I'd been looking forward to the encounter for years. The Great Predictor was famous—world famous. He was legendary for the number of his predictions, and for their amazing accuracy. Many people had relied on those predictions, and always with profit. And he could tell you the most amazing stuff—this, that, and the other thing.

What intrigued me the most, however, was how bizarre were some of the predictions. "Tomorrow you will be in two places at once" was one. "On Wednesday an event will occur for which there is no cause" was another.

How could such things be? I was captivated by the strangeness of these prognostications. Could such weird things really come to pass? That's why I had been so anxious to meet the Predictor. For years I had looked forward to finally getting to know him.

At long last I was getting my wish. I was twenty years old, and I was thrilled. I was sitting in a classroom, in college, on the first day of a course called Introduction to Quantum Mechanics.

Quantum mechanics is one of the glories of our age. The theory lies at the very heart of modern society. I once saw an estimate that a full third of our economy involves products based on it. Much of science involves it too, not to mention a good fraction of all the Nobel prizes awarded over the past century. Quantum mechanics is one of our most valuable forecasters, and its forecasts always turn out right. It has immeasurably altered our conception of the natural world. It is a triumph.

But it is not an unalloyed triumph. As the years have passed my initial admiration for quantum mechanics has become mixed with a certain confusion. I have never felt comfortable with the theory.

Let me introduce myself. In college and graduate school I studied physics—but when it came time to start doing research, I turned to astrophysics. That is the field in which I have worked for my entire career. But astrophysics is really just a branch of physics, so it was not so great a shift as all that.

And throughout my career I have maintained my early fascination with quantum mechanics. Somehow, I never felt that I really understood the theory. It always sat lodged in the back of my mind—enigmatic, mysterious, enticing. Over and over again, I found myself thinking that someday I really ought to go back and figure it all out, and finally put all those early juvenile confusions to rest.

For truth to tell, in college I never felt fully comfortable in most of the courses I was taking. Always I felt that yes, I was doing the homework, and yes I was getting by decently on the tests—but no, I did not fully comprehend those words, those formulas and equations and vast principles that my teachers were so confidently espousing, and that I was so dutifully memorizing. Throughout it all I kept telling myself that in the long run I would understand things. For the time being, however, memorization and practice solving problems would carry me through.

And I was right. Memorization and practice solving problems did carry me through. And more than that: in the long run, I did figure things out, and I came to some degree of understanding of all the various subjects I was studying.

All but one. All but quantum mechanics. That subject I never figured out.

Many years later, when I was in the middle of my career, I encountered a colleague who was as fascinated—and as confused—by the theory as I. As the years rolled by we kept discussing the issue, first casually, then more seriously. We formed a discussion group of like-minded colleagues. We organized a conference. And ultimately, we wrote a book on the theory's mysteries.

From the outset we knew that one of the book's chapters was going to be on something called Bell's Theorem. To be honest I found myself dreading getting to work on that particular topic. While I had never felt comfortable with quantum mechanics in general, Bell's Theorem was a topic that I felt positively unnerved by. Over and over again I had tried to master it, and over and over again I had failed. I actually recall wishing at one point that we could skip the whole damn thing.

In the long run we bit the bullet and sat down and worked out some sort of understanding of John Bell's celebrated discovery. We wrote that chapter, and we wrote the rest of the book, and it was published.

Nobody objected to what we had written in that chapter. No colleagues ever pointed out any errors within it. So, I told myself, we must have gotten it right. We must have actually figured out Bell's Theorem.

Skip forward many years. Time passed: my attention turned to other things. But as the years rolled by, I noticed an old, familiar sensation—the sensation, nibbling quietly at the back of my thoughts, that something was wrong, that something was still eluding me. And one day, I looked at my face in the mirror—this is literally true—and I spoke aloud. "Greenstein," I said to my reflection, "you were just kidding yourself, weren't you? You never really understood Bell's Theorem at all, did you?"

It was time to confess, and I did confess: in writing that chapter I had simply repeated the strategy that had proved so successful in college. I had said the right words and written down the right formulas—but I had not understood them.

"Time to get going," I told my reflection.

And I did. This book is the result.

John Bell's famous theorem had been meant to answer a specific question. It will take me several chapters even to describe the question he set out to address, and to set it in proper context. Suffice it to say here that Bell's question involves some of the deepest issues that human thought may address—issues involving the ultimate nature of being. All in all, it is an unusual situation. Physics is good at telling us how to fly to the moon, how to control magnetic fields or build a better clock. But such weighty matters as metaphysics? That's another matter.

People usually think that metaphysics—the study of existence and the ultimate nature of reality—is a purely philosophical subject. But Bell's Theorem showed that experiments could be performed that would tell us something about it. Thus a new field of study has come into being: not just physics, not even experimental physics, but experimental metaphysics. In the third section of this book I will describe it to you.

The experiments Bell suggested have been performed. The results are astonishing. I know of no easy way to briefly summarize the significance of those experiments, and the impact they have. But one thing I do know: they are revolutionary.

The Great Predictor can do so many different things. The range of his forecasts is astonishing. Quantum theory predicts the rate of radioactive decay.

It decrees that two hydrogen atoms will combine with an oxygen atom to form a water molecule, and it tells you the structure of that molecule and the energy released when it forms. It says that copper should be an electrical conductor but rubber an insulator. It predicts the structure of atoms—it predicts the very existence of atoms. It tells us this and that and the other thing. We have been immeasurably enriched by paying attention to these predictions.

If the Great Predictor were not so useful I wouldn't be so interested in him. And I wouldn't be so interested were not his forecasts so invariably correct. But they are correct. Not once in history has he ever been wrong.

To appreciate how remarkable this is, compare our Great Predictor with some other, lesser predictors. We have many in our society. An investor in the stock market forecasts how the market will behave. The weather bureau forecasts the weather. News media predict elections. Are they always right? Do they succeed in predicting the future in each and every situation? Of course not! In fact they do only somewhat better than the rest of us.

But what can we say of an investor who is correct more often than we? We say that she knows some things that the rest of us do not know. We say that she knows something about the innermost secret plans of corporations, regulatory agencies, and other investors. And since the weather bureau does not do all that badly we say that it knows a bit about the vagaries of wind, intrusions of high pressure, and shifts in humidity. The media know something about the opinions of voters. We say that the investor and the weather bureau and the media have to some degree succeeded in piercing the veil of appearances, and they have perceived something of an underlying truth that is hidden to the rest of us.

And the Great Predictor: what is the reality that he perceives? What are the truths that only he can see?

We physicists have a term for those truths: we call them "hidden variables." They are "variables" because they could have one value or another—an electron could be here or there, an atom could have this energy or that. And they are "hidden" because we do not see them: they are hidden from our gaze. "Hidden variables" is physicist-speak for what is actually going on: the real physical situation that we do not perceive, but that the Great Predictor apparently does—the reality about which he seems to know so much.

There's that word again: "reality." Metaphysics. I am starting to describe the background to John Bell's wonderful discovery. And, not to put too fine a point on it, the very question of whether hidden variables exist is the whole point of this book.

You might think it is all very obvious ... but if there is anything quantum theory has taught us, it is that nothing about the microworld is obvious.

It took me a long time to write this book. The reason is that I wrote it to put my thoughts in order—but those thoughts refused to settle down. They skittered around madly. I kept trying to understand the situation and failing. I would go through the proof of John Bell's wonderful theorem, not just once but over and over again—but I would end up as mystified as before. My problem wasn't the mathematics: it was what the mathematics meant. And when I asked myself what it meant ... why, my mind would just go blank.

That was a signal. I know myself well enough to realize that if I find it hard to even think about something, it is a message that there is some enormous gap in my understanding. Somewhere, something was missing from my thoughts. But what?

By now I know the answer to that question. By now I know that all along I had been operating in two ways at once. On the one hand, I was thinking in the normal way: the automobile is right *there* and it is going *that way* at *such-and-such* a speed. And on the other hand, I was thinking in terms of quantum mechanics. And what I now realize was that all along I had been operating in both modes at the same time. I was moving seamlessly and smoothly from one sphere of thought to the other. And most important of all: this moving from one to the other was unconscious.

If something is unconscious it just might cause you trouble. That, I ultimately came to realize, was what had been giving me so much grief for so very long.

In the pages that follow, I will invite you, the reader, to think along with me in the first of these two ways of approaching the microworld—the normal way, the non-quantum-mechanical way. I'm not doing this to be nasty. I'm doing it because this is how our minds naturally work. Not only that: it is how scientists approach their work—every scientist: biologists and geologists and chemists and, indeed, even physicists before quantum mechanics came along. It seems to be the right way to think. It is certainly the way I myself used to think. I'll go further: it seems to be the only way we *can* think.

There's only one problem: the new science of experimental metaphysics has shown that this way of thinking is wrong.

**Figure 1.1**
A typical tabletop quantum experiment. These experiments are not particularly spectacular to look at—but their results can be earthshaking. (This one is an experiment by David Hall of Amherst College, on a quantum phenomenon known as a Bose–Einstein condensate.) Photo by George Greenstein.

In the years that saw the creation of quantum mechanics, the theory's founders battled over the philosophical issues it raised. Albert Einstein, in particular, never accepted the theory—a theory he helped to create, and for which he won the Nobel Prize. For decades the arguments Einstein raised remained unresolved, until John Bell's famous discovery put them in a new and unexpected light. A series of experiments was performed—and I want to recount these experiments—that shed an extraordinary new light on the nature of the microworld.

The equipment involved in these experiments is not so very spectacular—nothing like the orbiting Hubble Space Telescope or the mighty Large Hadron Collider. Nowadays you can do a metaphysical experiment with gadgets that would fit on a tabletop. Anyone visiting such an experiment would come away pretty much unimpressed.

But while the experiments don't look so impressive, their results are. They are earthshaking. For in truth we have never before encountered anything like the revolution in thought that the new science of experimental metaphysics is forcing on us. It has led us into a realm unlike anything that has ever come before. We have known for decades that the world of the quantum was strange. But not until John Bell came along did we realize just how strange it is.

This book is about that strangeness. And it is about an argument that, although it began as a matter of abstract philosophy, by now has blossomed into what promises to be a multimillion-dollar industry, an industry based on quantum strangeness.

So there's a lot to talk about. Maybe it's time to get going.

# Background to Bell

# 2 Silence

If we are interested in the ultimate nature of reality, our Great Predictor is the one to talk to. After all, he obviously knows so much. Before Bell's Theorem came along, I would have said that I would dearly love him to tell me what he knows.

The problem would have been, though, that the Great Predictor doesn't talk very much. The Predictor is *reticent*. There are questions he never answers. If I ask him what will come to pass, he will reply with the utmost specificity. But if I ask for more he falls silent. Quantum theory makes predictions all right—but it does no more.

As an analogy, suppose that the Predictor tells me that tomorrow I will be in two places at once. And lo and behold, when tomorrow evening rolls around, I realize with a start that I do indeed have a vivid memory of having lunched with a friend at noon—and also a memory of having participated in a noontime pick-up basketball game. Of course my memory might well be mistaken, a delusion. But no! I ask my friend and he confirms having lunched with me, and all my basketball buddies vividly remember our game.

I'd better figure this out. So I approach the Predictor and ask him, "How is it possible to be in two places at once?"

The Predictor makes no reply. He refuses to answer my question.

Here's another analogy. There is a tree. It's autumn, the time that leaves turn color and fall to the ground. The Predictor says, "Next week half the tree's leaves will fall, while the other half will remain on the tree. The week after that, half the remaining leaves will fall. And so on."

I wait and watch. I find that indeed he had been correct. But now I find myself wondering, for many weeks into the autumn I notice that there are still a few isolated leaves clinging to the branches of the tree. They are resisting the buffeting of the winds. But why? What is the difference between the leaves?

Are those still on the tree more hardy, and those that fell more fragile? Do some have thick stems and others thinner? I climb the tree. I rummage around in the pile of leaves at its base. I find no differences. The leaves are all the same.

I go to the Predictor for enlightenment. I ask him, "Why did some leaves hang on for longer than others?"

Again the Predictor makes no reply. Again he refuses to respond to my question. He just sits there.

I ask again. I point to a particular leaf. "When will this one fall?" No response.

These are analogies, of course. The first is a translation into everyday terms of famous quantum-mechanical experiments that demonstrate that electrons can be in two places at once, as demonstrated by a phenomenon known as interference. Interference is a property of waves—but electrons are not waves, they are particles, and quantum theory refuses to explain how a particle can do such a thing. And the second analogy is of the radioactive decay of a nucleus, in which quantum theory correctly predicts the rate of decay, but refuses to explain why one nucleus decays sooner than another.

Don't focus on what quantum mechanics does. Focus on what it avoids doing. The theory steadfastly refuses to speak of many things. An electron can be emitted *here* and detected *there*, but the theory cannot describe the path the electron took. It tells us that an atom can have many different energies at the same time, but it does not tell us how this may be possible. It says that a particle can spin—indeed, that it must spin—but in no particular direction until it is observed. It tells us that events in the atomic realm occur randomly, but it fails to describe their causes. The theory deals only in probabilities, and it never gives explicit descriptions of events—first *this* happened, and then *that*. It never explains why an event occurred.

This refusal of the theory to respond to certain questions, this inability to give explanations for its predictions, to describe what happened, and to express certain things, deeply puzzled the theory's creators. And it has deeply puzzled many physicists ever since.

Return to our examples of everyday predictors. The investor, if she is in a talkative mood, might be willing to tell us what she knows about financial reality. A scientist in the weather bureau would be willing to bend your ear for hours on what he knows of atmospheric conditions. You can learn a lot about the mood of the public by talking to a media executive.

But you will never learn anything by asking the Great Predictor about the real physical situation that we do not perceive, but of which he seems to know so much.

There is a problem with the analogy I just used of the tree in autumn. The problem is that in my analogy the leaves were all the same. But real leaves are not all the same. If one stays up on its branch for longer than the others, there is a reason. Its stem might have been thicker, or maybe it was protected from the wind. But in the world of quantum mechanics there is no such reason, for radioactive nuclei are all alike—absolutely alike. And yet they behave differently.

So my analogy was flawed. And why did I not use a better one? Because there isn't any. In the world of our experience identical objects behave identically. And if two objects behave differently, it is because they only seem to be identical—were we to look more closely, we would eventually spot the difference. But for nuclei there is no difference.

There is a lesson in all this. It is that nothing in the normal world of daily experience prepares us for quantum mechanics. The microworld is alien—absolutely alien. If there is anything I have learned about the world of the quantum, it is that normal thinking simply does not apply.

So a brief warning. Throughout this book, I will often be using analogies. I will be doing this because I want to place the strange, unfamiliar world of quantum mechanics into a comfortable context. But some of these analogies are going to be misleading. I will try to warn you when they are, but it's going to be an awkward situation.

So be it.

And while I am at it, I should warn you of something else. Even now, so many years after the creation of quantum mechanics, physicists keep on arguing about it. There is still a profound disagreement among researchers about how to understand what it is telling us about the world. There have even been alternative theories proposed, designed to replace standard quantum theory with something else. Some of these can be thought of as re-interpretations of the theory, and some are outright modifications.[a]

a. The most prominent of these are the "pilot wave" theory developed by Bohm, the "spontaneous collapse theory" of Ghirardi, Rimini, and Weber, and the "many worlds" theory of Everett.

I want to warn you that everything I am going to say in this book refers to the standard version of the theory. Which is to say, I am completely ignoring these alternative approaches. Why do I do this? Because that is what most physicists do. None of these alternatives has attracted the amount of attention that the standard theory has attracted.

It is actually an unfortunate situation, for there is much to recommend each one of these other approaches. They are all worthy of more attention than they have received. As I am sure you will appreciate as this book goes along, standard quantum theory is utterly strange, and in the long run one of these alternative approaches may well prove to be a better way of dealing with the mysteries of the microworld. If so it will be that one that comes to attract the lion's share of physicists' attention.

But as of now, they lie on the outskirts. So in this book I will confine myself to the standard approach.

Here is something that I wish I could tell you: that back when I was a student, I simply could not reconcile myself to the Great Predictor's silence. That in those days I wanted to grab him by the lapels and shake him to and fro. That I wished I could yell at him to "speak! Say something! Explain yourself!"

But to be honest, I cannot really tell you this. Yes, I was sometimes irritated at the Great Predictor's silence. But as I have already mentioned, mostly I was irritated at myself. Irritated and even perhaps ashamed. Ashamed that I was so dense. Ashamed that I was Just Not Getting It. Ashamed that I was not understanding what my professor was trying to teach me. That professor certainly seemed pretty sure of himself, striding so confidently back and forth in front of the blackboard as he filled it with all that weird stuff. And, glancing sideways at my fellow students, I could not help but feel that they also seemed pretty sure of themselves. Was I the only one so confused? It was true in most of my courses.

Of course, I would never reveal such weakness in public. So I put on a brave face and soldiered on, writing in my notebook with a bored and superior mien. Who knows? Perhaps I even managed to fool myself.

For truth to tell, it is hard being a student. There is so much you have to learn. Everything is unfamiliar, and often it is foreign to your customary way of thinking. (I will illustrate this a little in chapter 6.) Before the school year began you had felt pretty sure of things—but now you are out of your depth, in new and uncharted territory.

It is true of every form of learning. Some time ago I decided that I needed to improve my tennis game. So I took a few lessons. The serve was something I found particularly hard. To this day I remember vividly all the contortions I was putting my body through in my efforts. I found myself twisting this way and that, bending into the most bizarre poses. Everything I was trying to do felt unnatural and awkward. Meanwhile my instructor was utterly graceful and at home as he demonstrated the proper form.

I now believe that the same is true of all learning. Tennis, quantum theory—it is all the same. To the newcomer it is alien and uncomfortable, and one's self-confidence can be undermined. In the long run, as the new material sinks in and is internalized, the sense of solidity and confidence returns. But in the first stages the whole situation can be very bruising to the ego.

And so it was with me. I found myself floundering during my initial exposure to quantum mechanics. And while this was going on, I simply had no energy for anything more than learning the material. It would be pleasing to me to be able to tell you now that, even as a student, I had questioned the very foundations of what I was struggling to learn. But that is not really true. The questions, I would tell myself, could wait.

But not always.

Every so often, I would approach my professor after class and try to speak about these mysteries. I wanted to ask him how an object could be in two places at once, or how something could happen without a cause. And although he was being polite, I could tell that my questions did not really interest the professor.

And more than that: I had a faint but unmistakable feeling that he regarded my questions as juvenile. "Kids," I would imagine him thinking. "You've got to love them—aren't they great? But in the long run this guy Greenstein will have to grow up."

Even then, at the very beginning, I had encountered a stigma—the faint but all-pervasive sense that I'd better not spend too much time asking such questions. I had encountered a second kind of silence. It was a silence that had paralyzed the field for decades.

# 3 Half a Theory?

This reticence is unique to quantum mechanics. No other theory of physics is so reticent. Newton's theory of gravitation speaks of the solar system as sort of a gigantic clock, a smoothly functioning machine. Statistical mechanics describes a gas as a swarm of particles, rapidly zooming about every which way. Electromagnetism tells of space filled with invisible fields. Each of these theories gives us a vivid picture of the world. But quantum theory tells us nothing of the sort: it leaves us no picture at all.

The refusal of the theory to give a picture of reality, to respond to certain questions, and its seeming inability to describe any mechanisms lying behind its predictions, deeply disturbed the theory's creators. One of those creators was Albert Einstein. He was so disturbed that he decided that there was something wrong with the theory. To him no respectable theory of physics should be so speechless.

Is there something wrong with quantum mechanics? It seems so limited! Is it *too* limited? Is there something wrong with it? Maybe it is not a theory but half a theory.

For surely there must be some way to penetrate beyond the silence of the theory! Surely something is missing from quantum mechanics. Could it be that it is only a partial theory, that underlying it there is a deeper understanding, a fuller and more complete explanation of the workings of the microworld? In this view quantum mechanics is only a start, and we need to search for a better theory that will speak of all those things that the Predictor fails to address. This new theory, if we could find it, would clearly describe the invisible reality—the actual physical situation, the real state of affairs—of which we would dearly love to learn.

**Figure 3.1**
Albert Einstein. Although he was one of the creators of quantum mechanics, he never accepted it. Over and over again, Einstein argued that the theory was incomplete because it failed to describe subatomic reality. The arguments that Einstein advanced are the fertile soil from which grew the discoveries described in this book. Photo courtesy of the American Institute of Physics Emilio Segre Visual Archives.

Does this view make sense to you? If you feel that it does make sense, then you believe in hidden variables. And it made sense to many people before Bell's Theorem and the experiments that it inspired came along.

In particular, it made sense to Albert Einstein. In his view quantum mechanics, successful though it may be, was *incomplete*. The theory was perfectly good so far as it went, he felt—but "so far as it went" was not good enough for him. Einstein wanted something more: a complete theory.

Is quantum theory incomplete? Or are there are reasons to think that it is all that we will ever get?

On the one hand, if we are going to revise quantum mechanics to make it into a complete theory, the task won't be easy. It won't be a matter of making a few tweaks here and there. The very structure of the theory, the way in which it is formulated, is antithetical to the goals of such a revision. If we are looking for a more talkative Predictor, it's an utterly different person that we're after. And if we do decide to go looking for a new one, we'd better remember that she's got some pretty big shoes to fill. Quantum mechanics as it stands is that good.

Furthermore, it seems inconceivable that such a Predictor—a talkative Predictor, one who describes what is really going on—could ever be found. What sort of description could there be of a particle existing in many places at once, of an atom with a broad range of energies, of an object most definitely spinning on an axis but of that axis having no particular direction until it is observed? The more we probe the quantum world, the more we learn that its properties simply cannot be described in anything like the ordinary manner.

Many of the theory's creators felt that the refusal of quantum theory to provide a detailed account of the workings of the microworld is not a *problem* but a *discovery*. It is not a limitation but an advance. In this view the theory is profoundly right in confining itself to doing only what can be done, and avoiding what cannot. These people insist that the question "What is the reality that the Predictor perceives so clearly?" is misguided—that there simply is no such thing. They insist that the idea of a reality that can be explained is a naive notion that we need to outgrow, on a par with another intuitively obvious concept that turned out to be wrong—that of a flat Earth beneath our feet.

Einstein, on the other hand, thought that these people were just spouting nonsense. He set out to prove them wrong. And it was the arguments that Einstein brought to bear that led in the long run to Bell's famous theorem, a theorem that, in a supreme irony, almost certainly showed that Einstein had been wrong.

# 4 The Solvay Battles

The issue around which Einstein chose to launch his attack involved one of the cornerstones of quantum theory: Werner Heisenberg's famous uncertainty principle.

The inability to give us a picture of a real physical situation (aka hidden variables) is built into the very structure of quantum mechanics. Perhaps the strongest evidence for this is provided by the uncertainty principle. This principle, formulated in 1927, deals with a strange limitation of the language of quantum theory: it cannot speak of certain *pairs* of properties. One example is the specification of the position and velocity[a] of an electron. Using the language of quantum mechanics we can write down a description of an electron being at some definite place—but that description will specify that the electron can have any velocity at all. Conversely, we can elect to write down a quantum-mechanical statement of the fact that the electron is moving at a definite velocity—but that description will specify that it can be at any location at all.

Are we just being stupid? Maybe with a little more work we could cook up a quantum-mechanical description of an electron with a perfectly definite position and velocity. Unfortunately, no matter how hard we try we find ourselves unable to find such a specification—and indeed, it can be shown that the mathematical structure of quantum theory is such as to prohibit such descriptions.[b] My Predictor does not even talk our language.

---

a. Technically the theory speaks of momentum, but momentum is just the electron's velocity multiplied by its mass so the distinction doesn't matter.

b. Years later a variant theory was proposed by David Bohm that evaded this problem. (It is briefly discussed in chapter 14.) This theory does not, however, succeed in providing the sort of mechanistic description of the microworld that Einstein was looking for.

But perhaps we have been misunderstanding the nature of the electron. Perhaps the electron is just not the sort of thing that *has* a definite location. Could it be a little fuzzy, or have a fluctuating shape, so that it is impossible to specify with perfect accuracy its position? Maybe an electron is less like a tiny, point-like particle and more like a region of bad weather. Things might be pretty stormy across the Northeast—but a more precise specification of the bad weather's location is simply impossible.

But this will not do, for quantum theory is perfectly capable of describing an electron located at a precisely specified place. It is only *pairs* of properties that cannot be simultaneously described. Such pairs are termed "complementary." Furthermore, it is only some pairs that are complementary: the theory is perfectly capable of simultaneously specifying the position and mass of an electron, for instance.

Is this a fatal limitation of quantum mechanics? Or is it some kind of insight into the very nature of the microworld? Is the uncertainty principle a problem or a discovery? Is it an expression of the fact that quantum mechanics is incomplete, a mere half-theory that must be supplanted by a fuller theory? Or is it a profound insight into metaphysics and the nature of reality?

Einstein took the first view. Another of the creators of quantum mechanics, Niels Bohr, took the second. Einstein wanted a more talkative Predictor; Bohr thought that this desire was naive. They battled it out for years.

Two memorable interchanges took place at historic "Solvay" conferences in the years just following the creation of quantum theory. Founded by the Belgian industrialist Ernest Solvay in 1911, these meetings take place in Brussels and have been devoted to fundamental issues in physics and chemistry. Meeting at irregular intervals, they continue to this day.

At the Solvay Conference in 1927—the very year in which Heisenberg enunciated his uncertainty principle—Einstein came up with a "thought experiment" that he felt revealed a way to circumvent it. Such thought experiments do not need to be performed: they are mental exercises designed to bring out certain elements of a situation, just as a novelist might place characters in a particular state of affairs to watch what they do. Einstein invented a series of steps designed to reveal to the experimenter two complementary quantities that, according to Heisenberg's principle, could not be so revealed.

**Figure 4.1**
Niels Bohr. Also one of the creators of quantum mechanics, Bohr argued that Einstein's search for a more complete theory—one that would describe microscopic reality—was misguided. Indeed, Bohr argued, the refusal of quantum theory to do so was not a problem but a discovery—a profound philosophical insight. Photo courtesy of the American Institute of Physics Emilio Segre Visual Archives.

A participant has given a first-hand description of what happened. Each day

> Einstein came down to breakfast and expressed his misgivings about the new quantum theory, every time [he] had invented some beautiful [thought] experiment from which one saw that [the theory] did not work. ... Pauli and Heisenberg, who were there, did not pay much attention, "Ah well, it will be all right, it will be all right." Bohr, on the other hand, reflected on it with care and in the evening, at dinner, we were all together and he cleared up the matter in detail.[1]

Three years later, at the next conference, Einstein arrived armed with a second thought experiment.

> It was quite a shock for Bohr—he did not see the solution at once. During the whole evening he was extremely unhappy, going from one to the other, trying to persuade them that it couldn't be true, that it would be the end of physics if Einstein were right; but he couldn't produce any refutation. I shall never forget the sight of the two antagonists leaving [the meeting], Einstein a tall majestic figure, walking quietly, with a somewhat ironic smile, and Bohr trotting near him, very excited.[2]

This time it took all night for Bohr to discover the error in Einstein's reasoning.

Both times Bohr had succeeded in refuting Einstein's arguments. But Einstein remained unconvinced nevertheless. He wrote to Schrödinger: "The soothing philosophy—or religion?—of [complementarity due to] Heisenberg–Bohr is so cleverly concocted that for the present it offers the believers a soft resting pillow from which they are not easily chased away. Let us therefore let them rest. ... This religion does damned little for me."[3]

Over the years the two battled it out. But their scientific disagreements never became personal. Indeed, Einstein held great affection for Bohr. Writing to a friend: "Bohr was here, and I am as much in love with him as you are. He is like an extremely sensitive child who moves around the world in a sort of trance."[4]

It cannot be said that the great debates between Einstein and Bohr ever reached a definite resolution. Rather they just seemed to peter out. Some people continued paying attention to the question, but by and large the mainstream did not.

Maybe it was a matter of simple exhaustion. Einstein and Bohr had wrestled over the matter without reaching a conclusion: why go over the same ground yet again?

**Figure 4.2**
Bohr and Einstein ... in the midst of a furious battle? Although they disagreed profoundly, their disagreements were never personal. In fact, they had deep affection and respect for one another. Photo courtesy of the American Institute of Physics Emilio Segre Visual Archives.

Or maybe it was like the famous paradox of Zeno: in order to go from here to there, first you need to cover half the distance, then half the remaining distance, and so on. This seems to prove that motion is impossible. But do I care? Not at all: I can't think of a satisfactory rebuttal, but that doesn't stop me from walking across the room. Similarly, maybe you don't have to know what quantum theory means in order to use it.

At any rate, thinking about such issues was just "not done" in those days. To many people it seemed a bit unprofessional, maybe even juvenile. Grownups did not waste time doing such things. Somehow, admitting to an interest in the nature of reality felt like admitting to a fascination with ESP or reincarnation.

On the one hand, the problem seemed to have no practical consequences. Whichever side you took made not the slightest difference to the conduct of your research. Accustomed to playing the hardball of theories that made specific, testable predictions, of conducting experiments that yielded detailed, verifiable results, stewing over such matters had been striking many people as just a bit fluffy.

It is also a matter of the technology available. In the great debates over the creation of quantum theory, the thought experiments of Einstein and others were just that: thought experiments. They could not actually be conducted. It was technically impossible. And in science, experiment and observation are paramount. Pure thought is all very well, but it gets you only so far.

The astonishing thing about the battle over the adequacy of quantum theory is that the battle simply seemed to vanish for many years. In my own experience, I can testify that when I was a student studying quantum mechanics not a single professor so much as mentioned the enigmatic nature of the theory, the mysteries surrounding its interpretation, or the great debates that had so animated Bohr and Einstein. The same is true of every textbook I ever read. Indeed, it was not until 1985 that a single graduate-level textbook so much as mentioned Bell's Theorem—more than two decades after he had discovered it. Undergraduate texts took even longer to get around to the subject.[5] And if our professors and our textbooks did not refer to the subject ... why then, we students did not either. The subject was out of bounds.

The Science sections of bookstores nowadays are crammed with popular expositions of the mysteries of quantum mechanics. This has not always been so. For more than two generations following the close of World War II virtually no such books appeared discussing these issues.[6] At the other end of the publishing spectrum, the *Physical Review*, one of the premier scientific journals in the world, had for years an explicit policy of refusing to publish any paper on such matters that did not explicitly make new quantitative experimental predictions.[7] Most of the great debates among the theory's founders would have remained forever unpublished under such a guideline.

So for decades the questions were relegated to the margins.

Einstein was trying to show that quantum mechanics was incomplete—a mere half-theory—and that it should be supplanted by a more complete one that would describe in detail the hidden reality of the microworld. So far he had failed. But he had not given up. And the next step he took led directly to Bell's wonderful discovery into the nature of quantum reality.

# 5 Spin

The simplest exposition I know of what came next involves electron spin.

Every rotating object possesses an axis of spin, represented as an imaginary arrow. Normally you can fully specify the orientation of that arrow: the spindle of a top, for instance, might tilt off to the right by so-and-so many degrees from the vertical.

Notice that in giving this specification I have made use of two different reference directions: one running right-and-left and one up-and-down. Fewer reference directions would not be enough to fully describe the orientation of the arrow: it would not be sufficient to say that the spin made such-and-such an angle to the vertical, had I neglected to also mention that it leaned right instead of left.

But quantum mechanics speaks a language all its own, and it is a strangely limited one. That language possesses no means of giving both specifications. Quantum mechanics can describe the fact that an electron's spin is up as opposed to down—but in such a configuration the theory is incapable of telling us whether it leans right or left. Alternatively, the language of quantum mechanics can express the fact that an electron's spin points to the right, but it cannot then specify the vertical component. If quantum theory is a language, it is an impoverished one, incapable of expressing many things.

Perhaps we should just try harder. Maybe with a little more work we could cook up a quantum-mechanical description of an electron with its spin arrow pointing in a definite direction. Unfortunately, no matter how hard we try we find ourselves unable to find such a specification—and indeed, it can be shown that the mathematical structure of quantum theory is such as to prohibit such descriptions. It is the Heisenberg uncertainty principle all over again.

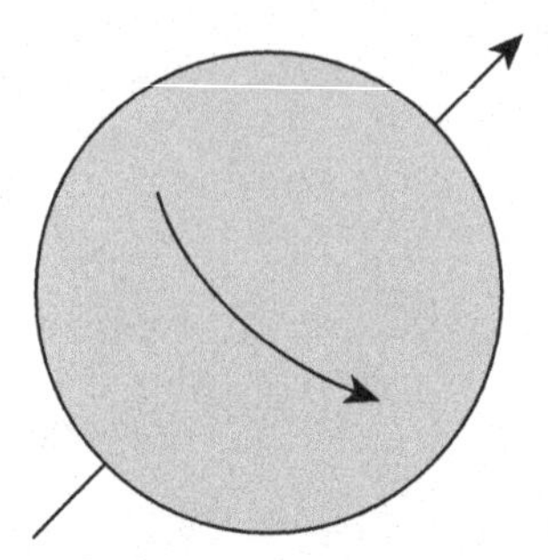

**Figure 5.1**
A spinning object and its axis of spin.

Sometimes things are even more ambiguous than that. There are quantum-mechanical states in which even the component of spin along a single reference direction is undefined. Imagine an electron gun—a device that shoots out electrons. (These guns used to be very common: they were constituents of television sets before the advent of flat-screen technology.) Suppose that such a gun produces electrons in such a strangely ambiguous state. It fires one off in the direction of an experimenter. That experimenter is equipped with a detector that measures the component of the electron's spin along a particular reference direction. What will it find? Will the electron's spin turn out to be this way or that? There is no way to know. Nothing in the quantum mechanics of such a state predicts the result the detector will get.

All the theory gives is the probability of a particular result. It is precisely this refusal to go further that is so frustrating about the theory. It seems to avoid all the interesting questions. Would you like to find a description of what is going on in the experiment? Do you find yourself beset by the urge to make up for quantum theory's lack? Do you want to posit some property of the electron flying toward the detector that explains the result that it gets? If so, then you are not alone. In the old days many people agreed with you. Einstein agreed with you.

Indeed, you may be feeling that the whole thing is trivial. Perhaps you feel that you already know what this property is. Perhaps you carry in your mind's eye an image—an image of a tiny speck hurtling away from the electron gun, aimed precisely at the detector. Perhaps in your imagination you lean forward to gaze closely at this speck—and you notice that it is spinning. Spinning about an axis that you can see in your mind's eye. An axis that points in a perfectly definite direction.

I too find it hard to resist this image. After all, it is what figure 5.1 shows. But that image is not provided by quantum theory. It is provided by a lifetime of experience in the large-scale world, a world that does not partake of the slippery, endlessly ambiguous nature of the microworld, a world that may be entirely inappropriate to this new zone of experience.

Indeed, quantum mechanics has no place for the image. It lies utterly outside the theory. It belongs to that world of which my Predictor refuses to speak—a world of real objects, of actual physical situations. A world of hidden variables. A world in which Einstein believed, but Bohr did not. A world whose very existence is in doubt.

For make no mistake: that apparently trivial figure carries with it a profound assumption about the very nature of reality. It is an assumption that I find almost impossible to question, but that I am being forced to question as I replay in my mind that long-ago battle over the completeness of quantum theory. An assumption about metaphysics. An assumption that John Bell realized could be tested by experiment.

# 6 An Impoverished Language?

What *does* quantum mechanics say about spin? I like to think of quantum theory as a kind of language. It is a strange, abbreviated tongue. Nothing in it seems to correspond to anything that we have ever experienced, and it is utterly incapable of expressing many things. Perhaps this is not surprising: it was designed to fit the strange new world of submicroscopic entities such as electrons and photons, not normal life.

One entity in this new form of discourse is something known as a matrix. It describes the electron's spin. A matrix like this

$$\begin{pmatrix} 1 \\ 0 \end{pmatrix} \text{ vertical}$$

**Figure 6.1**

represents an electron whose spin lies upward along the vertical direction. A matrix with the "zero" on top and the "one" on the bottom represents the opposite case of a spin pointing downward against the vertical. Similar matrices describe other directions—horizontal, say, or tilted. We can add such matrices, each multiplied by a number from which can be calculated the probability that a detector will yield the corresponding result.

All in all, it is a peculiar language, and one far removed from normal experience. This is what was giving me so much trouble when I was first introduced to it as a student. Nothing in this language corresponds to our customary, everyday notion of spin. And if we search through quantum theory, looking for the slightest mention of that notion, we will not find it. We will find no rotating bodies, no axes pointing off in definite directions. Rather, what we find are matrices—utterly foreign, and utterly abstract.

If you are of a certain frame of mind, such a way of thinking will be good enough for you. Mathematicians, for instance, are perfectly comfortable speaking in terms of abstractions with no counterpart in experience. (I once met a mathematician who mentioned that she was studying non-Euclidian geometry. "What a joy! It is so delightfully concrete!") For example, a mathematician is just as happy speaking of a negative number as a positive one. Maybe you are too. But should you be? We all know what four apples look like—but what do minus four apples look like?

The difficulty is that our normal way of thinking about a situation brings with it a mental picture of what is going on—but abstract thinking like this does not. Suppose, for instance, we ask "If there are ten apples in a bag, and four are removed, how many will be left?" Normally we would subtract to find the answer. But if you are of a mathematical bent of mind you might want to play with a second way of answering that question: that of *adding* a *negative number* of apples to the original ten. You will get the right answer. Notice, however, that this alternate way of dong the calculation fails to carry with it any such picture. Nobody has ever tossed a negative number of apples into a bag. Not even a mathematician.

In the same way, the language of quantum mechanics fails to carry with it a description of what might really be going on. Nothing in the matrices that constitute the quantum-mechanical description of spin corresponds to the sort of intuitive picture of figure 5.1. Do you find this a defect? Are you looking for more than matrices? If so, then you are looking for something that goes beyond quantum mechanics, something not contained within the theory and that encapsulates our normal conception of spin, our normal metaphysical assumption about reality.

Which is to say that you are looking for a theory better than quantum mechanics. You believe that quantum mechanics is incomplete. You are saying that it is half a theory, and it is not enough for you.

I don't want to give the wrong impression here. Many areas of physics use abstract mathematics. You will find the square root of minus one all over the place in the theory of electromagnetism, and non-Euclidian geometry in relativity. The point is that, in all these other areas, the math is about something that can be pictured.

In chapter 2 I spoke of the "reticence" of the Great Predictor, and of his "refusal" to answer my questions. In truth it is not really a question of reticence or refusal. Rather it is a question of his being *incapable* of speaking. The language that the Predictor speaks simply has no means of expressing all those things that constitute the reality concerning which Einstein wanted him to speak.

But maybe you—and Einstein—are wrong in your wish. For notice that, for all its abstractness, quantum mechanics makes perfectly definite predictions about the results of experiments. It predicts the probability that such-and-such a measurement will yield such-and-such a result. And if I make that measurement, I find that the prediction was correct. Isn't that enough? The Great Predictor got it right yet again: why isn't that good enough? If a theory predicts what I will find when I do something, is that not everything I should want of a theory?

There are times when I think that the language of quantum mechanics is not impoverished at all. To the contrary, it is perfectly adapted to the microworld—a realm utterly different than that of everyday experience, a realm in which our old ways of thinking are no longer appropriate. As with the uncertainty principle, perhaps the strange new language that the Predictor speaks is not a *problem* but a *discovery*—a profound insight into the very nature of physical reality.

I understand that Eskimos have many different words for snow. Quantum mechanics is like a language that has *no* words for snow. And the question is whether this is a failing of that language. Maybe not! Perhaps we should recognize it as a positive benefit of this way of speaking. For after all—maybe there is no such thing as snow. Maybe snow is like Santa Claus: a fiction that we are so accustomed to that we have mistaken it for the truth.

In this view, the image in figure 5.1 and the desire we may feel for a fuller description of what is really going on are instances of a naive, outmoded way of thinking. They are relics from an earlier time, from an earlier metaphysics, a metaphysics not yet informed by the wonderful discovery that is quantum theory. Better we should grow up.

# 7 The EPR Paradox

The year 1935 witnessed Einstein's most powerful salvo in his battle against Bohr. Referred to as "the EPR Paradox," this attack set the stage for positively endless arguments, disagreements, confusion, and fascination on the part of physicists for decades—and it directly led to John Bell's discovery. Indeed, Bell's Theorem is a direct extension of the EPR scenario.

The paradox appeared in a physics journal in the form of a brief paper bearing the forbidding—and ungrammatical—title "Can Quantum-Mechanical Description of Physical Reality Be Considered Complete?" Written together with Boris Podolsky and Nathan Rosen—hence the paper's acronym, EPR—it is one of the most historic scientific papers ever written. I can testify from personal experience that it is also one of the hardest to understand. I have pored over it for years, and I cannot say that I comprehend it in any depth. To me reading the EPR paper is a bit like wrestling with difficult poetry, or with prose lifted from some obscure Journal of Heavy Thought. It contains one of the most widely quoted—and, to me at least, incomprehensible—sentences of any scientific paper in any field:

> If, without in any way disturbing a system, we can predict with certainty (i.e., with probability equal to unity) the value of a physical quantity, then there exists an element of physical reality corresponding to this physical quantity.[1]

Succeeding work has clarified their argument. The simplest exposition I know of involves spin.

In their paper, Einstein, Podolsky, and Rosen provided an argument that shows that quantum spin was just like normal spin, and that it pointed in a perfectly definite direction. Since the language of quantum theory was incapable of expressing this fact, they argued, quantum theory was incomplete. It did not express all there was to be known. They claimed to have proven that the Great Predictor might be perfectly good at making predictions—but that he was utterly blind to the underlying reality about which he was so prescient. A new predictor was urgently required.

They did this by inventing yet another thought experiment, one designed to measure a spin's components along both the vertical and horizontal reference directions—again, just what the uncertainty principle forbids. Their experiment involves a sort of modified electron gun—a device that produces not one but two electrons. The device sits in the middle of a room. At opposite ends of the room are two experimenters. Let's give them names: Alice and Bob, say. The device in the middle shoots one of its particles toward Alice and the other toward Bob. As for the experimenters, each has a detector, and the two detectors are oriented the same way: both are vertical, say, or both horizontal. Each detector reveals whether the spin of the electron it observes points along this reference direction or against it.

The electrons produced by the emitting device are in something known as an "entangled state." Such a state has the property that the two detectors always yield opposite results. If one detector finds a spin pointing along its reference direction, the other is sure to find a spin pointing against it. We might liken such an entangled state to an angry couple. They are furious with each other, and they disagree constantly. If the husband is in the mood for seafood the wife will shudder at the very thought. If she wants to play golf, he wants to lie in the backyard. But it's not just a matter of his liking seafood, or of her feeling active—because he might just as well have hated seafood, and she might have been feeling lazy. Indeed, the only thing each of them really wants is, not this or that, but a fight. They want to disagree. So too with the EPR entangled state: the two detectors always disagree.[a]

In the EPR scenario Alice and Bob orient their detectors along the vertical direction, the button on the emitting device is pressed, and it shoots out such an entangled pair of electrons. Imagine that Alice is slightly closer to the device than Bob. She performs an observation: is her electron's spin up or down? Whatever she gets, she can predict the result that Bob will get: the opposite. And if Bob then performs his measurement, they will find that she had been right.

a. This is another instance of a poor analogy of which I warned earlier. In my analogy the husband and wife are able to disagree because each knows what the other wants. But in reality they do not know (I will explain why we are so sure of this in chapter 12). So even though they do indeed disagree, that disagreement is a mystery.

By observing what happens at her detector, Alice has found out something about the spin of her electron. But she has also done more: she has found out something about the spin of Bob's. And she did this without ever touching it. Einstein, Podolsky, and Rosen argue that *Bob's electron must have had this property even before Alice had made her measurement.*

For after all, they say, Alice and Bob's detectors were far away from each other, and hers could not possibly have influenced what was about to happen at his.

(As an aside, let me direct your attention to the forgoing sentence. Does it make sense to you? It certainly did to Einstein, Podolsky, and Rosen ... and yet, as we will see, it contains an assumption—an assumption that seems utterly obvious, and yet that in the long run will turn out to be false. Let us give that assumption a name: *locality*. We will return to the subject of locality in chapter 14. But for now let us proceed.)

Now Alice and Bob twist their detectors about so that they lie along the horizontal direction. They repeat the procedure. As before, Alice can now determine not just the vertical, but also the horizontal component of the spin of Bob's electron. So, say EPR, both the vertical and the horizontal components of the spin of his particle exist—in contradiction to the uncertainty principle.

The argument seems to make a lot of sense. If the experiment is repeated over and over again, Alice always succeeds in predicting the result of Bob's measurement. But why? Why does Bob's detector keep on getting the predicted result? What influences his detector's behavior when a particle arrives? The only possibility seems to be the particle itself. This is the only thing that can "tell Bob's detector what to do." If this seems reasonable to you, reflect that we are now speaking of some attribute, carried by this particle, that influences Bob's detector. Does this strike you as a correct way to think? If so, then you believe in hidden variables. You believe that the spin of the particle had a perfectly definite value even before Bob observed it.

It might appear to be a trivial argument. As an analogy, imagine that you are holding a coin. Cut it carefully along its flat plane, so that you end up with two half-coins: one is heads and the other tails. The half-coins are analogous to the spins of the entangled particles, and of the states of mind of the angry couple.

Take two envelopes: slip the "heads" half-coin into one and the "tails" into the other. Shuffle the envelopes and then mail them off. One goes to

Alice and the other one to Bob. Alice gets to her mailbox first: when she opens her envelope she finds herself able to predict with certainty what Bob will find when he opens his.

It seems almost churlish to point out that, in attributing a "headness" or "tailness" to the content of an envelope, we are doing precisely what quantum theory is incapable of doing. Those faces are hidden variables. They are properties of the half-coins. The very existence of such a property was what Einstein and Bohr were fighting over. It is what everyone has been fighting over since the creation of the theory.

They are what the Great Predictor will not speak about. And if he will not speak about them ... then maybe he is not so very Great after all.

A colleague of Bohr has given us a description of the EPR paper's effect:

> This onslaught came down upon us as a bolt from the blue. Its effect on Bohr was remarkable. ... A new worry could not come at a less propitious time. Yet as soon as Bohr had heard my report of Einstein's argument, everything else was abandoned: we had to clear up such a misunderstanding at once. We should reply by taking up the same example and showing the right way to speak about it. In great excitement, Bohr immediately started dictating to me the outline of such a reply. Very soon, however, he became hesitant. "No, this won't do, we must try all over again ... we must make it quite clear. ..." So it went on for a while, with growing wonder at the unexpected subtlety of the argument. Now and then, he would turn to me: "What *can* they mean? Do *you* understand it?" There would follow some inconclusive exegesis. Clearly, we were further from the mark than we first thought. Eventually, he broke off with the familiar remark that he must sleep on it.[2]

In the days that followed Bohr developed a counterargument to that of E, P, and R. It was rapidly published. Very few colleagues whom I have consulted claim to understand this paper.

Indeed, much of Bohr's thought seems to be obscure to the point of incomprehensibility. One contemporary physicist has grumbled, "Whatever the merits of Bohr's approach, it did not really facilitate answering awkward questions; it was better at giving verbally dexterous accounts of why they could not be answered."[3] And John Bell has commented, "Bohr was inconsistent, unclear, willfully obscure, and right. Einstein was consistent, clear, down-to-earth, and wrong."[4] Finally, a quote from the Nobel Prize winning physicist Murray Gell-Mann: "Niels Bohr brainwashed a whole generation of theorists into thinking that the job [of understanding quantum mechanics] was done 50 years ago." [5]

# 8 Hidden Variables

Figure 5.1 is precisely what we might imagine my Great Predictor sees as he peers beyond the veil of appearances to see the underlying reality—that reality of which he so adamantly refuses to speak. Vividly obvious in that figure is something that quantum theory fails to give us: an actual situation. We see in this figure a full specification of the direction to which the spin points. We can even measure that direction, and come up with a definite number: so-and-so many degrees off to the right.

That number is the hidden variable describing the electron's spin. We call it a "variable" because it could have one value or another—fourteen degrees, for instance, or maybe one hundred and ten. And it is "hidden" because it is tucked away, hidden from the theory's gaze. Quantum mechanics cannot give us the value of a hidden variable. Indeed, it seems to have no place for them within its way of doing things.

The problem goes beyond spin. It infects everything that quantum theory addresses. Consider as another example the matter of radioactive decay, in which an atomic nucleus spontaneously breaks apart into pieces. Such nuclei are found to possess a certain half-life—the length of time during which half of them decay. There is an isotope of radium, for instance, that has a half-life of a bit more than eleven days. Start out with a chunk of pure radium: if you come back eleven days later, half the atoms will have decayed. If you wait yet another eleven days, half the remaining ones will be gone.

Quantum theory can make predictions about this half-life. But what the theory cannot tell us is when any given radium nucleus will decay. Imagine that two such nuclei lie before you. They are identical: nevertheless, if you come back a couple of weeks later, you might find that one of them has decayed while the other is still intact. But why? What is the difference between the two, which allowed one to survive longer than the other? The theory has nothing to say on the matter.

In speaking of radioactive decay I have already used the analogy of leaves falling from a tree (chapter 2). Here is another analogy. Suppose that the weather bureau has predicted a 50 percent chance of rain on Monday—but that when Monday rolls around, you find that it fails to rain. Suppose further that again on Tuesday the prediction is for a 50 percent chance of rain, and that again Tuesday remains sunny. So it might go for several days until, on Saturday (for which the prediction is still a 50 percent chance of rain), the rain finally comes.

What was different about Saturday? We do not know, but we have some suspicions. Maybe it was a sudden incursion of cooler air from Canada, which the weather bureau had not foreseen. Or maybe it was a lower-than-usual air pressure coming up from the south. Weather is complicated, after all, so we forgive the weather bureau.

We may not know the reason for Saturday's rain, but we are positive that this reason exists. That reason is a hidden variable. If we ever managed to figure out why it rained on Saturday, it would no longer be hidden. It would be a "seen variable"—a known variable.

In terms of radioactive decay, if there is a reason, quantum theory does not give it to us. Just as the theory has no way within its language to give a precise specification of the direction of spin, it has no way to predict when any given nucleus will decay. If you are looking for such a reason, you are looking for a hidden variable. And if you are looking for a hidden variable, you are looking for a more complete theory than quantum mechanics. A better theory. A loquacious Predictor, one willing to speak more openly.

John Bell's Theorem dealt with the question of hidden variables. Do they exist? Is there a reality that my Great Predictor sees, but refuses to speak of? Is quantum mechanics a mere half-theory, destined to be replaced by a clearer view of reality? Or are hidden variables and the reality they represent a naive fantasy, a leftover from an earlier and outmoded picture of the world?

We are ready to turn to Bell's Theorem.

# Bell's Theorem

# 9 A Hidden Variable Theory

It was many years ago that I first encountered the Great Predictor.

I was thrilled to meet him. I'd been looking forward to the encounter for years. The Predictor was famous—world-famous. He was legendary for the number of his predictions, and for their amazing accuracy. Many people had relied on those predictions, and always with profit.

What intrigued me the most, however, was how bizarre were some of his predictions. "Tomorrow you will be in two places at once" was one. "On Wednesday an event will occur for which there is no cause" was another.

How could such things be? I was captivated by the strangeness of these prognostications. Could such weird things really come to pass? That's why I had been so anxious to meet the Predictor. For years I had looked forward to finally getting to know him.

At long last I was getting my wish. I was twenty years old, and I was thrilled. I was sitting in a classroom, in college, on the first day of a course called Introduction to Quantum Mechanics.

That was many long years ago. And throughout my career I have maintained my early fascination with quantum mechanics. Somehow, though, I never felt that I really understood the theory. It always sat lodged in the back of my mind—enigmatic, mysterious, enticing. Over and over again, I found myself thinking that someday I really ought to go back and figure it all out, and finally put all those early juvenile confusions to rest.

Part of that project was an effort to understand Bell's Theorem. To be honest I found myself dreading getting to work on that particular topic. While I had never felt comfortable with quantum mechanics in general, Bell's Theorem was a topic that I felt positively unnerved by. Over and over again I had tried to master it, and over and over again I had failed.

Eventually I did come to some sort of understanding of Bell's work. I recall feeling pretty pleased with things ... until the fateful day when I looked at my reflection in the mirror—and this is literally true—and I spoke aloud. "Greenstein," I said to my reflection, "you were just kidding yourself, weren't you? You never really understood Bell's Theorem at all, did you?"

"Time to get going," I told my reflection.

And I did.

What did I do? I read some books. I read some scientific articles. Among all my readings, one stood out: an article provocatively entitled "Is the Moon There When Nobody Looks?," whose explicit function was to make Bell's Theorem as accessible and as comprehensible as possible. I found it wonderful: immensely readable and immensely informative.[a] I read that article—not just once, but over and over again. It was helpful ... and yet, no matter how many times I returned to it, I still felt that something was eluding me. Some central insight, some clear understanding I still found out of reach.

Eventually I realized the error I had been making. I had been trying to adopt someone else's thoughts and make them my own. And then I realized that it was time for me to stop heading down this other person's path, and strike out by myself. It was time to take seriously the central mystery, and try to think about it in my own personal terms.

I decided to tackle the famous and intimidating EPR paradox.

At first glance it might be hard to see what the fuss over the EPR Paradox is all about. Why do people even call it a "paradox?" Perhaps you, the reader, feel this way. After all, is it not obvious what is going on in the EPR thought experiment? Is it not clear that the source of electrons, however it may work, is merely shooting out a pair of particles, one flying toward Alice and the other toward Bob, set spinning in opposite directions (figure 9.1)?

Figure 9.1 illustrates the underlying reality that we might think our Great Predictor sees, but of which he so adamantly refuses to speak. And isn't this simple picture all we need to explain the fact that Alice's and Bob's results disagree?

I decided to take seriously this picture, and see how well it dealt with the EPR scenario. I decided to see if I could create the very hidden-variable

a. You will find a reference to this article in the "Further Reading" appendix.

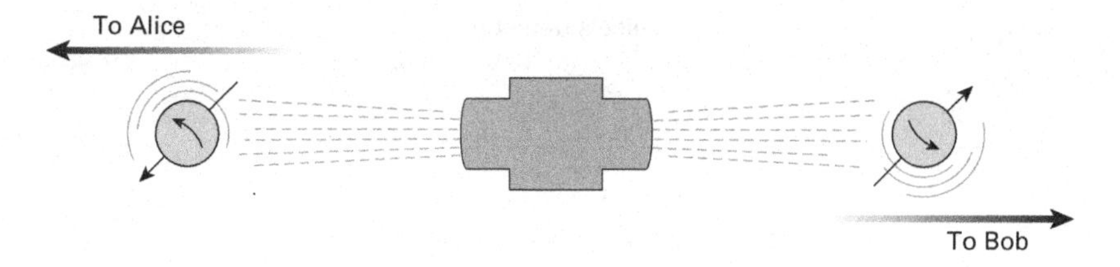


**Figure 9.1**
Naive view of the entangled state. It makes intuitive sense, but it will turn out to be wrong.

theory that Einstein sought, and Bohr declared impossible—a theory that made the same predictions as quantum theory, but that went further and described in full detail the workings of the microworld. I would create my own predictor—a new predictor, a loquacious predictor, one who was willing to speak up: a "Greater Predictor." Notice that in doing so I was doing just what quantum theory failed to do.[b]

How far could I go in this project? I was not so naive as to think that I was going to succeed. As you can guess, my project failed. I did not end up accomplishing what right then seemed so easy. So am I wasting your time as I lead you through this exercise? I am not—because the reason I failed will turn out to be very instructive indeed.

Before we begin, a brief word. If mathematics makes you nervous, just skip over the details of what follows and resume reading toward the end of this chapter. But perhaps you might want to resist this urge, and just read on. Maybe it won't be so very terrible after all!

A difficulty facing me at the outset was that a detector never told the precise direction a spin pointed. All it could reveal was whether the spin pointed more or less along or against the detector's reference direction. If Alice's detector found, let us say, *against*, the only thing I knew was that the spin of her electron must have one of the configurations illustrated by the dotted arrows in figure 9.2.[c]

---

b. The theory was going to make the same "locality" assumption that Einstein, Podolsky, and Rosen made in their famous EPR paper—an assumption so obvious that they did not even bother to discuss it, but one that we will need to devote a whole chapter to later on.

c. In what follows I am going to do a simplified analysis in which the spin arrow always lies in the plane of the page.

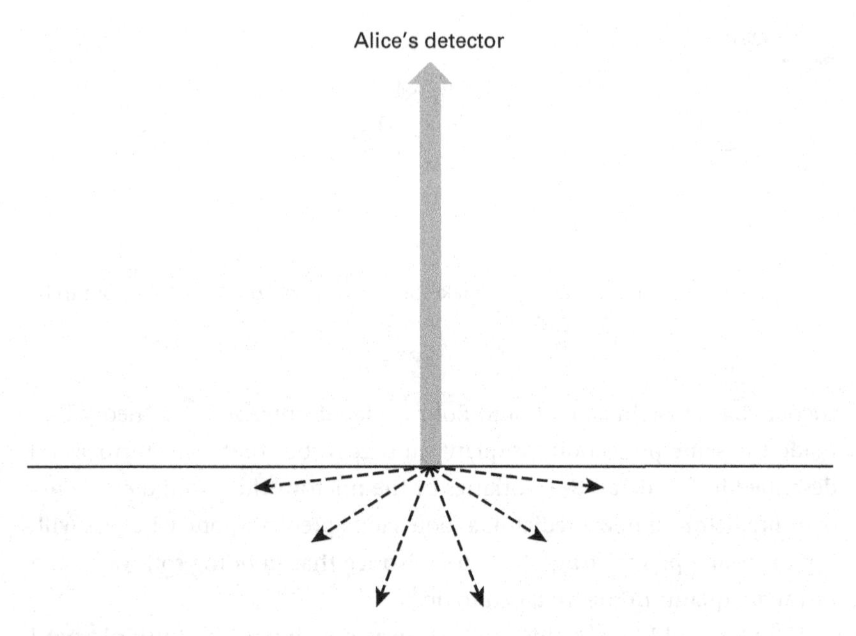


**Figure 9.2**
The spin axis of Alice's electron points along one of the indicated directions.

But I knew little more than that: any one of the dotted arrows might be the spin axis of the particle she had just detected. Her detector did not say which.

But there was one more thing I did know: since within my hidden-variable theory the spin of the particle heading toward Bob was opposite to that of Alice, its spin must have one of the configurations shown in figure 9.3.

And since Bob's detector lay in the same direction as Alice's, he was sure to find the spin of the particle heading toward him to lie *along* his detector's reference direction. So his measurements would invariably disagree with those of Alice. The electrons were behaving just like the angry couple I described in the EPR chapter.

This was a prediction of my hidden-variable theory—and it was just what the EPR thought experiment revealed. My theory was doing well so far: its predictions were the same as those of quantum mechanics. So maybe I had succeeded in my project of developing the hidden-variable theory that underlay quantum theory.

But only so far—and this was where the genius of John Bell came into play. Bell added a new twist to the EPR scenario. He asked what would happen were Bob to swing his detector about, so that it was no longer parallel with Alice's (figure 9.4).

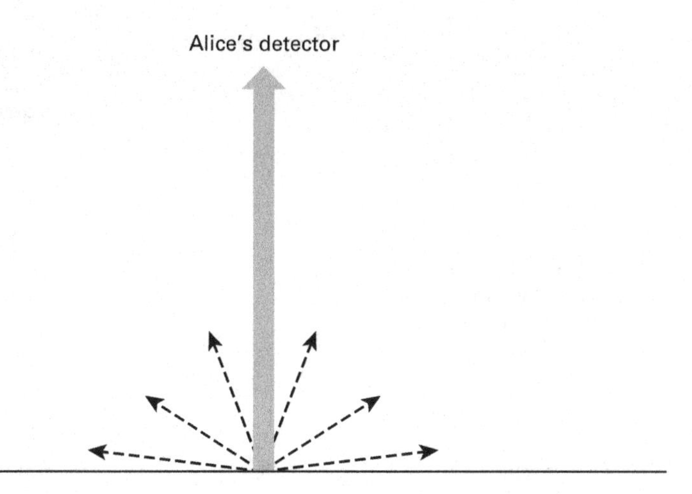


**Figure 9.3**
The spin axis of Bob's electron points along one of the indicated directions.

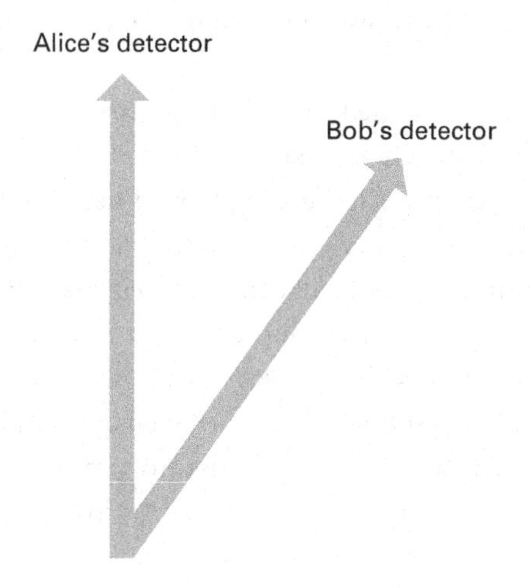


**Figure 9.4**
Bob's detector points in a different direction than Alice's.

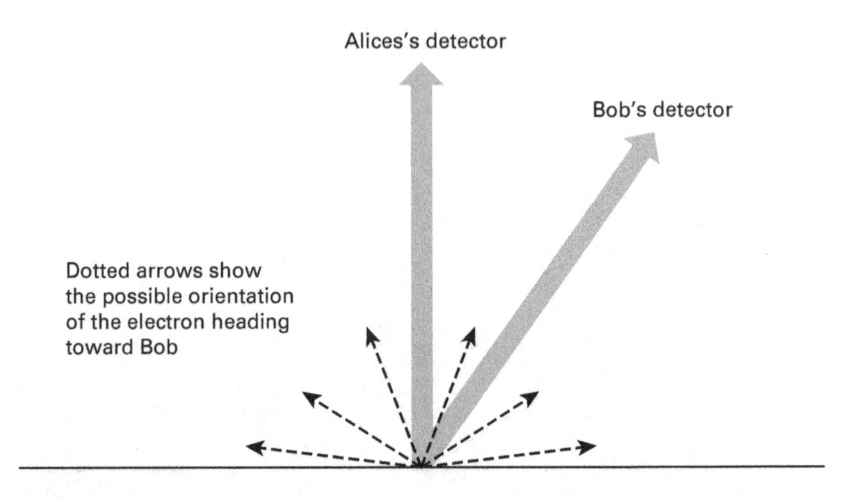


**Figure 9.5**
The electron that reaches Bob has one of the indicated spin axes.

Would the predictions of my hidden-variable theory still match those of quantum mechanics?

Suppose that Alice happened to find her electron's spin to be *against* the direction of her detector. Then the other electron's spin would lie *along* it. Of course, that electron was not heading toward her: it was heading toward Bob. When he measured it along his new direction, what result would he get? Would he continue to disagree with Alice so regularly? Or was the angry couple in our analogy starting to find at least a few areas of agreement? I needed to figure out *how many times Bob's and Alice's results disagreed.* And I needed to compare the result with that predicted by quantum theory.

There's no need for secrecy: I found that my prediction failed to match that of quantum mechanics. So I had failed in my quest to improve on quantum theory. Let us see how this comes about.

In my hidden-variable theory there were many possible orientations the spin of Bob's particle might have (figure 9.5). Were they *along* or *against* his detector's direction? Well, some were the one and some were the other (figure 9.6).

In figure 9.6 the possible spin directions that led to a disagreement between Bob and Alice were those that lay in the dashed arc. But not all of the possible spins lay within that arc! There was a "bite" taken out of it.

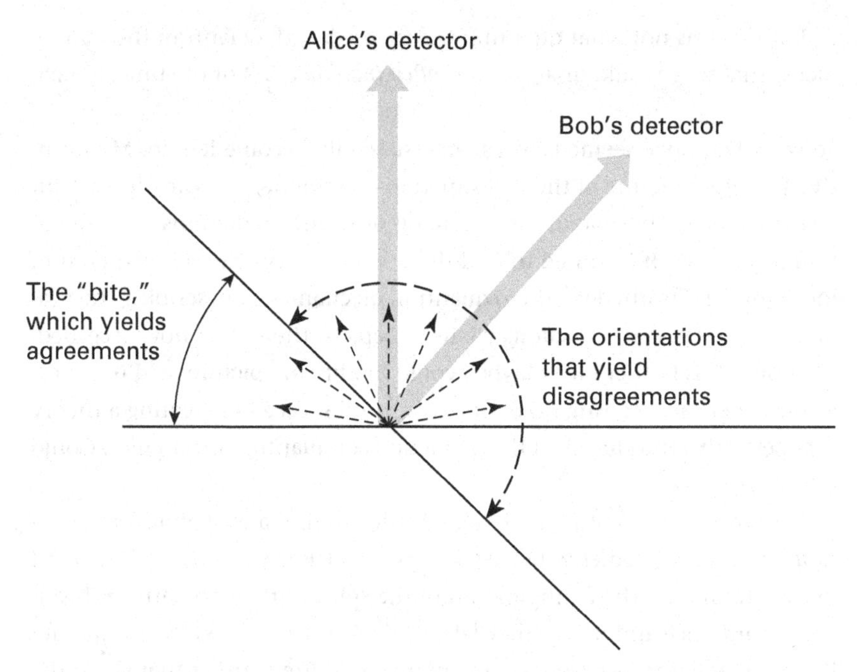


**Figure 9.6**
When do Bob's and Alice's measurements agree?

So there would be some agreements—those configurations in which Bob's arrow lay in the bite.

Just to be specific, I imagined that Bob's detector lay at an angle of 60° to Alice's. Then that dotted "bite" was also 60° in extent. Since the dashed arc's extent was 180°, the part of it which contained spins which disagreed with Alice was 180°–60° = 120° in extent. And this span was 120° / 180° = 2/3 of the full span.

Whew! But that was my result: of all the possible directions the spins may point, 2/3 of them led to a disagreement between Bob's measurement and Alice's. So now I had my hidden variable theory's prediction: when the detectors had been aligned, all measurements disagreed—but if they were tilted by 60°, only 2/3 of them did.[d]

---

d. I got this result based on Alice's having gotten the result *against.* Had she gotten *along,* the result turned out to be the same. It's not surprising, since the two situations are perfectly symmetrical.

But this was not what quantum theory predicted. Quantum theory predicted that they would disagree *more often than that*: 3/4 of the time, in fact.

So what had once seemed so easy had suddenly become less so. My intuitively obvious picture of the EPR situation—so simple, so clear—turned out to have failed. That picture was making different predictions than quantum mechanics. If I wanted to find the deeper theory underlying quantum mechanics, if I wanted to add to quantum mechanics by describing the hidden variables it so conspicuously fails to depict—then I had not succeeded.

Should this bother me? Maybe I could modify my picture, adding a new element here and getting rid of one there, and succeed in creating a theory that perfectly reproduced all the predictions of quantum mechanics. Could I do this?

I already knew what I had to do: I needed to find a way of *increasing* the number of disagreements between Alice's and Bob's results. To do this I needed to modify the configuration of the spins of the two particles heading toward Alice and Bob as they left the device that created them—modify it in such as a way as to reduce the number of dotted arrows that lay in the "bite" in figure 9.6. Was this so hard to arrange?

Not at all. Indeed, it was easy. I just had to rearrange the orientations of some of the electrons as they zipped off from the electron gun.

But not so fast! Bob might have oriented his detector, not 60° to the right, but 60° to the left. And in this case, that strategy would have exactly the wrong effect—not of increasing the number of disagreements, but of decreasing it.

The problem with this attempted fix was that Bob was free to orient his detector in any way at all, and whatever strategy I adopted was sure to work for only some of the orientations he might have chosen. No strategy would work for every possible choice. My theory could not be adjusted to yield the same behavior as quantum theory.

Before, I had likened the EPR state to an angry couple, intent on disagreeing with one another when asked the same question. Our new situation is more complicated: now each could be asked one of many different questions—sometimes the same, sometimes different, but questions they have no way of anticipating. In order to reproduce the predictions of quantum mechanics they would have to synchronize their replies, sometimes agreeing and sometimes disagreeing. But they can't: neither of them knows which question is asked of the other, nor does either know the reply given. So they have no way to synchronize their replies appropriately.

But they do it anyway.

So I was glad that I had gone through my little exercise. It had not really failed at all—for I had never honestly thought that I would succeed in creating a theory with which to challenge one of the greatest achievements of twentieth-century thought. Indeed, that had never been the point. The real point of the exercise had been to make clear to me just how enigmatic quantum mechanics really was. For at first it had all seemed so simple … and now I realized that it was not so simple at all. Indeed, suddenly it was profound. For now I realized that quantum mechanics was making a prediction for which there was no sensible explanation.

# 10 Bell's Theorem

Over the decades that saw the creation of quantum theory, the questions I have been discussing were central. People argued over them endlessly. I think it is fair to say, though, that the argument never reached a conclusion. Rather, as I have described earlier, it just petered out. People gave up talking about the subject.

But then John Bell came along.

In the previous chapter I tried to create my own hidden-variable theory, and I failed. I could not make it work. How about some other theory? Maybe I should abandon figure 9.1's simple picture, so intuitively obvious, of what is going on in the spin experiments, and try something else.

Bell's genius was to prove that the task is impossible. His theorem shows that no matter how hard I try, I will not be able to create a hidden-variable theory that agrees with all of the predictions of quantum theory. Yes, some of its predictions will agree—but there are sure to be others that don't. And the amazing thing about his theorem is that it doesn't deal with any particular hidden-variable theory at all. It deals with hidden-variable theories in general.

Bell's Theorem is concerned with an EPR configuration in which the two detectors are no longer parallel—just as I had done in my attempt. In such a configuration it no longer remains the case that the two detectors always get opposite results. Sometimes they do and sometimes they don't. Bell considered three possible configurations, in which the detectors were oriented in three possible ways. As I had done, for each of the three configurations he considered the fraction of times they disagreed. He was able to construct a specific mathematical expression involving these fractions. His thinking had nothing to do with quantum mechanics. As a matter of

**Figure 10.1**
John Bell lectures. His famous theorem showed that no local description of submicroscopic reality could make the same predictions as quantum mechanics. On the blackboard behind him can be seen a segment of this famous theorem (at the top). Photo courtesy of CERN.

fact, it had nothing to do with physics. It was pure logic: a matter of analyzing all the ways a random variable can be distributed.[a]

As an analogy, when you flip a fair coin it lands heads 50 percent of the time. An unfair coin, though, might be more likely to land heads than tails.

---

a. In this work, Bell made the same locality assumption as had Einstein, Podolsky, and Rosen in their famous EPR paper—an assumption so obvious that they did not even bother to discuss it, but one that we will need to devote a whole chapter to later on. So his theorem applied to *local* hidden variable theories only.

The degree to which the coin has been altered is a hidden variable, which you do not know. But you do know that if you take the fraction of times the coin lands heads, and add to it the fraction of times it lands tails, you will always get the number one. In an analogous way, Bell constructed a specific combination of the three fractions from the three experiments of his scenario. And he found a definite restriction involving this combination, which each and every hidden-variable theory must obey.

That was his result. But he then went on to do something more. Bell showed that quantum theory violated this restriction.

The conclusion? That quantum theory is different from every possible local hidden-variable theory. That quantum theory is something else.

John Bell has an almost mythical status among his peers. A colleague has written that he "was called the Oracle ... there was a certain aura about him."[1] Another has written of how "Bell's presence in a gathering raised the collective level of thinking, speaking and listening."[2] A third—a friend of mine—once recounted to me how he had happened to meet Bell by accident one day and how the encounter, brief as it was, had left my friend positively breathless. The praise of one's colleagues is the finest praise.

Bell was born in Belfast in 1928 to an impoverished family: his father was regularly in and out of work. He was one of four children. Schooling was an expense his family was able to meet only with difficulty. He began his university career not as a student but as a technician in a physics department. Members of the faculty, recognizing his talent and commitment, went out of their way to help, lending him books and allowing him to sit in on lectures. Ultimately he got an education, found employment as a physicist, and eventually moved to Switzerland and the giant particle accelerator at CERN, the European center for high-energy physics. There he worked on the quantum theory of fields and on accelerator design—and on the foundations of quantum mechanics. No one has probed those foundations more deeply.

In person Bell was graceful yet intense, and suffused with a quiet humor. These qualities shine through in his writings. These writings are graceful, passionate, powerful—and delightful. A few quotations will make this evident.

From an article of his on the mysteries of quantum theory: "The concept 'velocity of an electron' is now unproblematic only when not thought about."[3] And another: "The typical physicist feels that [these questions] have long been answered and that he will fully understand just how if ever he can spare twenty minutes to think about it."[4] From a review Bell had

**Figure 10.2**
John Bell and his wife Mary at dinner with friends. Mary Bell is also a physicist: indeed, the two Bells often collaborated. Photo © Renate Bertlmann.

written—and he is referring to himself here: "Like all authors of noncommissioned reviews, he thinks that he can restate the position with such clarity and simplicity that all previous discussions will be eclipsed."[5] And finally a quote from a letter to a colleague who had sent him a paper proposing a revised quantum theory: "I read with very great interest and admiration your paper.…Of course I will be happy to receive any reply from you. But I hold the right not to reply to letters to be the most fundamental of human freedoms."[6]

It was my good fortune to have met Bell. How that came about makes for an interesting tale.

In chapter 1 I referred to "a colleague who was as fascinated—and as confused—by the theory as I." That colleague's name is Arthur Zajonc (rhymes with "science"): his office was just down the hall from mine. As I wrote in chapter 1, we talked over the years, first casually and then more seriously. Ultimately these conversations led to a book. And at another point they led to a conference that Arthur and I organized. That was where I met Bell.

Here is what we wrote in our book about that conference.

> Conferences are the stuff of life to the working scientist. The lectures provided at conferences provide an in-depth view of the latest advances in the field: often these lectures are collected and published as a book, which stands as an invaluable summary of the current state of the art. Paradoxically, however, what participants often find most valuable in a conference is not these lectures. Rather, it is what happens in the nooks and crannies lying between the formal presentations: the brief conversation over coffee, the chance encounter in the hallway, the scientific argument that erupts over dinner. We decided to organize a conference that would consist of *nothing but* these informal chats.
>
> The meeting we envisaged was to be a week-long conversation, tightly focused on [the mysteries of quantum mechanics]. Attendance was to be kept low: in addition to the local physicists from the host institutions, only a limited number of the world's foremost workers in the field would be invited. Total immersion, we decided, was an important consideration: the conferees would sleep under the same roof and eat all their meals together. It was essential that the accommodations be comfortable and the meals tasty. We decided to hold it at Amherst College, our host institution.
>
> After eight months of planning and preparations, the conference began with a cocktail party on the back porch of what had, until recently, been a college residence hall. The school year had just ended: hard upon the heels of the departing students, a small army had descended on the building—cleaning up the mess, moving in new furniture, and putting fresh sheets on the beds. The sun shone brightly down, puffy white clouds marched across a sky of perfect blue, and the meeting's participants enthusiastically pumped one another's hands. Some had flown in from Europe or across the United States, others had driven for hours, and yet others had walked over from their offices. Although some collaborated regularly, others had not seen one another for years: indeed, a few had known each other only as names on scientific publications and were meeting for the first time. Suitcases stood around unattended as their owners, not bothering to carry them up to their rooms, fell deep into conversation.
>
> The next morning, after a gourmet breakfast, we gathered around a large table downstairs. One of the participants [it happened to be John Bell that first morning] stood up to deliver a brief talk. He had not spoken for five minutes before someone interrupted him with a comment. Someone else then chimed in with a comment on the comment—and we were off.

> For the rest of the week, participants gathered around the table to discuss some of the most fascinating and profound problems of modern physics. ... But the conference did not take place only around that table. It also took place out in the garden, and on the streets of Amherst as two or three participants would break away from the group and wander off for a stroll. It took place over meals, which were invariably huge and invariably delicious. The conversations would zip from topic to topic with astonishing rapidity—from the implications of a recent experiment to the standings of the New York Mets, from culinary delights to a possible new theorem four of the participants thought they had just discovered (those four skipped going out to a movie with the rest of us one evening, stayed up to all hours, and eventually decided that their promising new approach was probably no more than a dead end ...).
>
> At one point [during a picnic] Greenstein spotted John Bell and [another participant] off in a corner ... that afternoon they had squared off. For years Bell had probed with astonishing brilliance and depth the foundations of quantum theory, and he has argued that the theory is plagued by fundamental inadequacies. [The other participant], in turn, has argued with great subtlety that the mysteries of quantum mechanics have been widely exaggerated, and that in reality the theory poses difficulties no deeper than those raised by many other branches of physics. That afternoon their debate had risen to a passionate intensity.
>
> Greenstein grew perturbed. [The two] stood with heads together, isolated from the rest of the throng. Were they at it again? Had emotions risen so high that they had grown furious with one another? Greenstein sidled unobtrusively over to eavesdrop—and found them quietly comparing their cameras.[7]
>
> Tragically, Bell died, entirely unexpectedly, shortly after the conference.

As perhaps you can tell, Bell made an extraordinary impression on me. His presence was both gigantic and gentle. I felt that I was in the presence of someone who thought more deeply, more intensely, and more honestly than most—and who was at the same time among the most considerate people I have ever met.

Never have I encountered a person more committed to getting to the heart of a matter, and to a ruthless clarity and honesty. He was passionate in his argumentation. This was so even when he was an undergraduate. During his very first course in quantum mechanics its mysteries had disquieted him—as did what he regarded as too cavalier an attitude toward these mysteries on the part of the teacher. He remonstrated with that teacher. According to Bell's own testimony years later, he got into a heated argument, essentially accusing the professor—a man who had gone out of his way to assist him—of dishonesty.

And yet John Bell was one of the gentlest people I have ever met. He was considerate, respectful, gracious. He would fight over ideas, but the fights were never personal. You could disagree with him without angering him. And in some remarkable way, he could attack your ideas without attacking you. John Bell was a gentle battler.

Bell's work was pioneering. He showed the way to a new insight. His was the first step—and the first step is always the hardest one. You may have noticed that I have made no attempt to show the actual theorem Bell had proved. Rather, I have talked about it only in general terms. The reason is that his combination of mathematical quantities is something of a complicated mess, and the proof that they obey his restriction is more complicated still. But in the years following on Bell's work other people have proved theorems analogous to his, going beyond his work in various ways. In particular, a generalization of his work has been found that is so simple and straightforward that it is actually possible to give the proof in a nontechnical book. This "Bell-like theorem" is described in the appendix.

Bell's combination of mathematical quantities is, to me at least, quite strange. Nothing about it has any clear and simple interpretation, any immediately obvious intuitive significance. It just happens to be true that this particular combination of fractions happens to have the property he found. Indeed, my guess is that nobody would have found his result even slightly interesting—were it not for the amazing fact that quantum theory makes a different prediction.

How on earth did Bell do it? What led him to formulate this particular mathematical expression, and to ask whether it agreed with the predictions of quantum theory? I know of no writings in which Bell explained the train of thought that lead him to his theorem. I know of no one who asked him.

But I can hazard a guess as to what Bell might have replied to such a question. He might have said the same thing that every creative person would say: that it involved work—lots and lots of work, months and years of immersing himself in the problem, of wrestling with it day in and day out, living with it morning, noon, and night. He might have said that he used one approach and then another, trying this and trying that, groping about in the darkness—but a darkness that, as the effort wore on, seemed to be slowly lifting. He might have said that he sought to identify what was good about an approach that almost worked, and what was bad about one that came nowhere near working, and that he modified his various attempts accordingly. He might have said that the final result seemed to have come to him in a flash (but only after endless slogging). Or he might have said that it came to him gradually, incrementally, with agonizing slowness.

But most of all, I imagine, he might have leaned back in his chair and smiled, and said that in the last analysis he really did not know how he had done it.

# 11 Stigma

One day in his office at CERN Bell was visited by a young physicist who told Bell that he was gearing up to do an experiment based on Bell's Theorem.

"Do you have a permanent position?" was Bell's response.

It was a joke—but only something of a joke. Bell knew all too well, indeed from personal experience, that there was a stigma attached to doing such work. Bell's stature in the field was secure, but that of his visitor was not: the proposed experiment was to count toward his PhD dissertation work. Bell knew that the young man was taking a risk, for his experiment would take place amid a general atmosphere of indifference.

Not for eight years after his groundbreaking discovery was the first experiment conducted to test Bell's result. And not for seventeen years was the second set of experiments performed. That does not exactly add up to a burst of excited attention. The same pattern is found in the attention paid by other scientists to the article in which Bell announced his discovery. Most important discoveries in science are greeted by an immediate and intense burst of interest. In contrast, Bell's was greeted with a near-complete disinterest. Hardly anyone paid any notice. It was only slowly, over a period of more than a decade, that attention to his theorem slowly built up.

This indifference was part of a wider pattern. Perhaps my own testimony will be illuminating here. I was certainly aware, in a vague way, of Bell's result in those days. But I paid it little attention. Why? Had you been there to ask me, I would have said that it was simply because I was busy with other things.

And to be honest, this is a perfectly valid response. At every stage of one's life, there are a few things that one is doing, and there is a well-nigh infinite number of things that one is *not* doing. By now I find myself fascinated with Bell's work and its implications—but, by the same token, I am not working on the composition of the rings of Saturn, or the origin of the moon, or the nature of quasars. There are only so many hours in the day.

Looking back on my career, it seems to me that many times my choice of what to work on has been influenced by what everybody else was working on. If everybody in my field was fascinated by such-and-such an issue, then I was likely to be too. And usually that issue was indeed fascinating, and worthy of attention—and indeed, much fine work was being done on it, and many wonderful results obtained. Why turn one's back on all this good stuff? Why spend time way off in the boondocks, working on an issue that nobody else cared about?

But in the end I cannot say that I find any of this fully persuasive. None of it really accounts for the great silence that greeted Bell's work. There was a near-total lack of interest, a well-nigh universal shrug. Was it all just a matter of fashion, of the style of work people prefer to do? (The reader may be surprised to read of fashion playing a role in science. And of course it has no place if we are speaking of its dictating the results of scientific research: the physical universe is not a matter of opinion or taste. But it is another matter entirely if we are speaking of the choice of what research to conduct.)

Why did the great mystery, which so engrossed the founders of quantum mechanics, simply seem to pass out of fashion for so many years?

Part of the reason was John von Neumann.

Von Neumann was a mathematician, one of the most prestigious of his age. Born in Hungary to wealthy parents—his father, a banker, was elevated to the nobility—he was raised in an 18-room apartment on the top floor above the multigenerational family business. There, his parents realized that they had a child prodigy on their hands. By the age of six he was doing the sort of arithmetic that the rest of us would find hard on paper: he was doing it in his head. Within two years he had learned calculus. At 15 he began private studies with a prominent mathematician who found himself blinking back tears on encountering so prodigious a talent. By his late twenties he had published 32 major scientific papers, at the rate of nearly one a month.

This extraordinary genius was no recluse. He loved the good life. He appreciated food (his wife joked that he could count everything except calories) and drink and conversation. And he loved clothes. Invariably dressed formally, he was so elegant that one of his teachers inquired at his doctoral exam "Pray, who is the candidate's tailor?" He did his best work surrounded by cacophony, and at his office would regularly blast loud German march music out on his phonograph. He was a ghastly driver.

Von Neumann's fellow mathematicians were frankly astounded at his talent. One, whose lectures von Neumann attended as a student, said that he "was the only student I was ever afraid of." He made seminal contributions over an extraordinarily wide swath, ranging from the foundations of mathematics all the way up to game theory and economics. He worked on weapon design—he took part in the Manhattan Project, which developed the atomic bomb, and he contributed to the design of the hydrogen bomb.

And he worked on quantum theory.

In 1932 von Neumann published a book entitled *Mathematical Foundations of Quantum Mechanics*. It was a hugely influential work—a work of which he himself was quite proud. And in that work he proved—proved mathematically, proved rigorously, and without a shadow of a doubt—that hidden variables had no place in quantum mechanics.

My guess is that many physicists found themselves positively relieved at this result. If there is one thing that the ongoing argument between Bohr and Einstein proved, it is that the question of the interpretation of quantum mechanics is hard—very hard. And now one of the most famous mathematicians in the world had relieved them of the burden of carrying on the task. There were no hidden variables. Quantum mechanics was not half a theory. It was a full theory.

There was only one problem. Von Neumann's proof was only a "proof." It contained an error.

But the error was subtle and it eluded people for decades. During the intervening period, everyone thought that the problem had been solved. It was only years later that the error was discovered.[a]

So it is a complicated story. For decades physicists found themselves free to ignore the historical argument that had once raged over the nature of quantum reality. Since they felt they did not have to deal with it, they did not deal with it. So there was plenty of reason for these matters to be relegated to the sidelines.

---

a. The error was discovered in 1935 by Grete Hermann—and then again independently by Bell himself many years later.

But I believe there is more to the story than this. Because I believe it was not just a matter of indifference. It was a matter of scorn. I think that discussion of the subject was greeted in those days with an active antipathy. Look, for instance, at where Bell had chosen to publish his result. Avoiding all the mainstream journals, he had placed it in a new and obscure journal, one that promptly went out of business. Why did Bell hide his great discovery away like this? Why was he worried? I'd say he was worried because of the stigma. At the time, Bell was on sabbatical leave, and he apparently felt uncomfortable asking his host institution to pay the cost of publishing anything on so outlandish a topic. Most scientific journals support themselves financially by assessing authors a fee to cover the cost of publication: these fees can be quite steep. So he chose to publish in one of the few journals that did not levy these charges.

What had changed from the days of the Bohr–Einstein debates to this?

The historian of science David Kaiser attributes this shift in attitude to the cataclysm of World War II and the exigencies of the Cold War. Federal funding for physics shot up enormously during the war—by a factor of over 50 within a mere seven years. Such a flood of money was bound to have a transformative effect on the field.

The annihilation of Hiroshima and Nagasaki forever altered the status of physics in the eyes of the world—and in the eyes of physicists too. Once gentle souls, lost in their ivory towers, physicists suddenly found themselves acting more like hard-nosed captains of industry. Once they were philosophers, now they were warriors:

> Before the war, Einstein, Bohr, Heisenberg, and Schrödinger had held one model in mind for the aspiring physicist. A physicist should aim, above all, to be a *Kulturträger*—a bearer of culture—as comfortable reciting passages of Goethe's *Faust* from memory or admiring a Mozart sonata as jousting over the strange world of the quantum. The physicists who came of age during and after World War II crafted a rather different identity for themselves. Watching their mentors stride through the corridors of power, advising generals, lecturing politicians, and consulting for major industries, few sought to mimic the otherworldly, detached demeanor of the prewar days.[1]

It was not just a matter of self-image. Physicists were being enlisted as soldiers in the Cold War. People were urgently needed to combat Soviet domination in the arms race and the space race—highly trained people skilled in the arcana of the new physics. Enrollments in physics courses exploded. At the start of the World War the United States was turning out fewer than 200 physics PhDs per year—but by 1960 that number had more than tripled, and by 1970 it had passed 1,500. In such an atmosphere there was no need—and no time, and maybe even no stomach—to go into such charming stuff as the ultimate nature of reality. Tough people were wanted: can-do types who would roll up their sleeves, brush aside the niceties, and get down to cases. People with deadlines to meet and jobs to do.

And the world got what it wanted.

Experimental Metaphysics

# 12 Experimental ...

Does Bell's discovery show the hidden-variables idea to be worthless? Not at all! What it shows is that any local hidden-variables theory is going to be different from quantum mechanics. Maybe it is quantum mechanics that is worthless. For after all, suddenly we realize that we have not one theory but two: on the one hand quantum mechanics; and on the other hand something else, a hidden-variables picture of the world that does everything that quantum mechanics cannot do and explains the workings of the microworld. It is not a failure to realize that there can be no hidden-variables picture of the world underlying quantum theory. Perhaps it is a wonderful opportunity.

Because if you have two theories, you might want to ask—which one is right? An experiment can decide. An experiment in metaphysics.

John Clauser did the first experiment.

Clauser was born in California. He went to college at Caltech and then to Columbia University for graduate school. It was while he was at Columbia that he had read Bell's paper announcing his discovery.

> It was incredible to me. I didn't understand it or couldn't believe it. ... I thought "if I don't believe it, I should be able to give counterexamples" [to prove it wrong]....So I tried to and failed. I realized: this is the most amazing result I've ever seen in my life.[1]

He had never been happy with quantum theory's refusal to provide a picture of the physical world:

> I am not really a very good abstract mathematician or abstract thinker. Yes, I can conceptualize [quantum theory's mathematics]. I can work with it, I can sort of know what it is. But I can't really get intimate with it. I am really very much of a concrete thinker, and I really kind of need a model, or some way of visualizing something in physics.[2]

**Figure 12.1**
Clauser later in life. John Clauser performed the first pioneering experiments on Bell's Theorem. Bell had shown that we have not one but two theories: one is quantum mechanics, and the other is some theory that would fully describe the reality underlying quantum phenomena. Clauser realized that an experiment could be done that would tell us which was correct. His result favored quantum mechanics. Photo courtesy of John Clauser.

Perhaps it was the abstractness of the theory that bothered him. As a matter of fact, all abstractness bothered him:

> I had great problems all my life understanding the square root of minus one. In high school, I learned that the square root of minus one was called this little symbol i. Well, there is no such actual number ... it's an imaginary number. If you multiply i times i you get minus one. All right. But suppose I go to the store to buy 1+i candy bars. I could buy one candy bar, or one and a half candy bars, but I couldn't buy 1+i candy bars. But it's useful because it makes the equations work out, and once you play with it, the equations work out better that way. So then when I get to [college

people would say] "Well, it's a mathematical artifact. Don't worry about it. It just makes the equations look nicer." ... I was not very good at it; and didn't understand, didn't know why I was doing it. And I felt very uncomfortable with it. And once I felt uncomfortable with it, my brain kind of refused to do it.[3]

Clauser may have disliked abstraction—but he loved experiments. His father had been an engineer.

As I grew up, basically as a kid, I just would come in after school to his lab. We lived in the suburbs, and so I would do homework—I was supposed to be doing homework, but mostly what I would do is just sort of wander around the lab and gawk at all of the nifty laboratory equipment. And I kept thinking, "Wow, boy do these guys have fun toys. When I grow up I want to be a scientist so that I can play with neat toys like this."

I was an electronics whiz kid. [My dad] taught me some of the basics of electronics, and I just went off and built some of the earliest computers and the like. I built the world's first video games, and I actually won a whole bunch of prizes in the National Science Fair for doing this.[4]

Clauser is enamored of gadgets, and he loves to do experiments. He has lots of patents. One is for a device inspired by something he read in Tom Clancy's thriller *The Hunt for Red October*. Another is for a new kind of sail (he is an accomplished yachtsman, with many trophies). Clauser's garage has done double duty as a laboratory. A bookshelf at home is crammed with catalogs from scientific equipment supply houses. "Anything I need to make, if I don't have the pieces already, I look for it here. I can make anything."[5] "I've gotten pretty good at dumpster diving. ... If you are innovative and clever, it's amazing what you [can do]."[6]

As for quantum theory, nobody else seemed to share his misgivings about it. "I sat in my corner and tried to understand it myself. Nobody else talked my language."[7] He was alone as he stewed over what he felt to be the theory's deficiencies.

And the more he stewed, the more Clauser became convinced that quantum theory could not be the whole story. He decided that there had to be hidden variables. And so:

The Vietnam War dominated the political thoughts of my generation. Being a young student living in this era of revolutionary thinking, I naturally wanted to "shake the world." Since I already believed that hidden variables may indeed exist, I figured that this was obviously the crucial experiment for finally revealing their existence. But if they do exist, then quantum mechanics must be verifiably wrong here, with its error having gone undiscovered heretofore. ... To me, the

> possibility of actually experimentally discovering a flaw in quantum mechanics was mind-boggling.[8]

So he resolved to do an experiment.

Before he did that experiment, there were a few loose ends to tie up. One was that he needed a different version of Bell's Theorem. Bell had concerned himself with an ideal situation in which experiments get perfect results. Real experiments, on the other hand, are imperfect. In collaboration with colleagues, Clauser produced a new version of the theorem, one appropriate to such a situation.

> I submitted my thesis to Columbia, and I think there was like two weeks or so between submitting the thesis and the thesis defense, which was kind of a dead time. And so I just went up to Boston—actually it was to Wellesley and stayed in [one of my colleague's] house with him. And [the other colleague] came over pretty much every day, and we just sat there and took two weeks to hash the whole thing out.[9]

The other loose end was that he had gotten a job across the country, at the University of California at Berkeley. He needed to get from New York to California—and he needed to get his yacht out there too. So he resolved to go by sail.

> I had the job out [in California], and I had a boat [in New York City]. And originally, we were just going to sail the boat all the way to Galveston and put it on a truck there, and truck it across to LA and sail it up the coast to Berkeley. It turns out we ran into Hurricane Camille, so we got kind of stopped at Fort Lauderdale. We didn't save any extra mileage by doing this, but we had a lot of fun sailing down the coast. So every time we put into a port, I would get on the phone and [one of my colleagues] knew my schedule. And so basically he would send off his re-drafts to all of the various marinas in the next city where we put in, some of which I picked up, and some of which are probably still sitting there for all I know. While I was sailing, I would be writing furiously away and editing various things. And we'd get on the phone and chatter about various versions, and we'd keep swapping drafts. This continued all the way until I got to Berkeley, writing the paper, and then we finally submitted it, pretty much right as I arrived in Berkeley.[10]

If you want to do quantum theory, you need to think about things like the square root of minus one. If you want to do an experiment testing quantum theory, you need to think about things like sandwiches and cardboard.

I had a friend who worked with a particle accelerator. Every time they fired up the machine they needed to pump out all the air within it (you don't want any air molecules flying around: the particles you are trying to accelerate would bump into them). One day they turned on the vacuum pump and it just couldn't clear out all the air. It would pump and pump, but the pressure gauge always showed a faint residual pressure. Finally they got tired of waiting. They turned off the pump and opened up the accelerator.

Inside they found a half-eaten ham sandwich. Someone had inadvertently left the thing inside. It had been outgassing, the emitted volatiles spoiling the accelerator's vacuum. "It was pretty desiccated by the time we got to it," my friend allowed. "As if it had been freeze-dried."

Another friend once clapped a sheet of cardboard over the front end of a cutting-edge experiment he was doing—just to keep out the dust. As I recall he also used cardboard to fashion a small, meandering dam around the base of the experiment, to keep out any puddles of water that might form on his laboratory's floor.

The famous Cosmic Background Radiation, the faint glow left over from the Big Bang, was discovered by two scientists using a gigantic radio telescope. When they got ready to start up the telescope their first project was to get rid of a "milky white dielectric substance" they found coating its surface. The substance turned out to be bird droppings. Another radio astronomer I know used to deal with bird droppings by driving his car right up to his antenna and gently bumping it, to shake the stuff loose.

Clauser built his experiment with the help of a graduate student, Stuart Freedman. The device they built stood maybe waist high and it was about 10 feet long. Inside, an oven heated a chunk of calcium to the vaporization point. Individual atoms streamed out at several thousand miles per hour. They entered a chamber, where they were illuminated by light at a set of precisely calibrated wavelengths. The light induced the atoms to emit a pair of entangled photons (the experiment used photons rather than electrons). These were sent off in opposite directions. Plates of glass formed polarizers—the photon analog of the direction an electron spin detector would point. The photon detectors were immersed in an ultracold slush to improve their performance. Every so often motors would rotate the polarizers to a new configuration, as required by Bell's Theorem:

> At the end of each hundred-second counting cycle, the machine automatically paused and a sequencer (an old telephone relay that Clauser had rescued) would

order one or the other polarizer to turn 22.5 degrees in an orchestrated cacophony of domino-like noise and action, vivid in Clauser's mind thirty years later. "These big mama two-horsepower motors would crank over the "coffins" [holding the polarizers] and the teletype would clatter away," the paper tape pummeling, accordion-style, into a peach basket, spraying its chads across the floor, to the *ka-chunk* of the serial printer monitoring the quartz crystal that monitored the calcium beam.[11]

Two years to build the experiment. Two years of seemingly endless checking and re-checking, of improvising and fiddling and fussing over the details.

An experimenter has to fuss about all sorts of things. You can't take anything for granted. You have to understand every facet of your experiment. Students often don't understand this: they want to charge ahead, throw the apparatus together, turn it on and get a result. Their instructors can have a hard time of it slowing them down.

Clauser:

People always think you don't have the time to test everything. ... The truth is you don't *not* have the time. It's actually the time-saving way of doing it. It's hard, when you're eager to know what Nature's doing: you almost have to train yourself to be *anti*curious while you're building your hardware. People always want to slap it all together, turn it on, and *see what happens*. But for the first run, you can almost guarantee it's not going to work right.[12]

If you don't like this sort of thing, you don't want to become an experimentalist. Many people don't—me, for instance. Others do. I was chatting with a friend recently about the whole business of getting an experiment up and running. He was talking about how careful he had to be, how many errors there were to be chased down, and how slow was the progress. At one point he leaned back, gazed at the ceiling, and mused. "There are so many things in an experiment to worry over," he said. "So much to get under control. It is really—"

Before my friend finished speaking, I thought I knew what he was going to say. In my mind I finished his sentence for him. In my mind I had him saying something like "a royal pain in the ass." But, as it turned out, that was not what he said.

"Fascinating" was what he said. And he smiled again.

Clauser kept slogging. Throughout it all, he had finally found other people interested in his favorite topic, the mysteries of quantum mechanics. This was a wild and woolly group of physicist-hippies—it was, after all, the age of the counterculture and the antiwar movement—that called itself the Fundamental Fysiks Group. Its members would meet every Friday afternoon around a table at the University of California at Berkeley. There they engaged in a wide-ranging, free-form discussion of a breathtakingly wide variety of subjects. Topics ranged from the significance of Bell's Theorem to faster-than-light communication, from ESP to LSD.

Clauser joined the group—sort of: "Those guys were a bunch of nuts, really. ... But we kind of used that as a forum. The real physicists were over here in one corner, and all the kooks are in the other corner."[13] There were meetings at the Esalen Institute in Big Sur:

> The guy who was running this decided that quantum mechanics was related to this consciousness expansion, and would bring us down there. It was free for us, and there were hot sulfur baths that were there, and the rocks, and you'd go into these hot baths, all communal, with everybody naked, which I guess was part of the grand excitement. And then the hot water would sort of overflow the tubs and go cascading down the cliff into the Pacific Ocean. And so part of the highlight of every evening was a trip to the baths. And then during the day, we would sit around and talk about new aspects of quantum mechanics and the like, and how it was related to the great cosmic cockroach, or whatever. None of which I thought very much of, but what the heck?[14]

Eventually, Clauser and Freedman got their machine up and running. Once built, it ran for a total of two hundred hours spaced over several months. They got a result.

Their result disagreed with the local hidden-variables hypothesis, and it supported quantum mechanics.

Clauser's pioneering experiment had been conducted in the face of the scorn and antipathy of which I wrote in the previous chapter. And Clauser paid a price. Many colleagues felt little interest in his work, showed little interest in its results, and felt free to advise others of their opinion. He never got a position at a university or college—this in spite of glowing recommendations from prestigious senior colleagues. "I believe he shows promise of becoming one of the most important experimentalists of the next decade," wrote one. But it was to no avail: over the decades following on his experiment Clauser was forced to work in a research laboratory or on his own as an entrepreneur.

In 2002 Clauser wrote about his early experiences. His article rings with irritation at the reception his work had received:

> Most of the [subject] represented forbidden thinking for practicing physicists. Indeed, any open inquiry into the wonders and peculiarities of quantum

> mechanics ... was then virtually prohibited by the existence of various religious stigmas and social pressures, that taken together, amounted to an evangelical crusade against such thinking.[15]

Later on in his article Clauser writes of McCarthyism and relates it to

> a very powerful secondary stigma [that] began to develop within the physics community towards anybody who sacrilegiously was critical of quantum theory's fundamentals. ... The net impact of this stigma was that any physicist who openly criticized or even seriously questioned these foundations (or predictions) was immediately branded as a "quack."[16]

In the long run, Clauser's work has been widely recognized as a pioneering triumph—he received a prestigious award for it in 2010. But the long run was far off in the future while he was doing his experiment.

A quick aside.

It was when he was a graduate student at Columbia University in New York (working on a PhD thesis involving astrophysics) that Clauser became captivated by Bell's Theorem. It turns out that he and I are pretty much the same age—and I was working at a research institute just a few blocks down the street from Columbia. I remember meeting him. We did not get to know each other well. It was a matter of just a few encounters. But even now, decades after these encounters, memories stand out in my mind.

One was that Clauser lived, not in an apartment like everybody else in New York, but on a yacht—a yacht of his own, which he moored in some marina nearby. (Recall that, on finishing up at Columbia he had set out to sail all the way to California. He is a serious yachtsman.) Unusual enough. Another was that all he wanted to talk about was Bell's Theorem.

And there's one other memory. I recall what I said to him. "Bell's Theorem? Never heard of it. What is it?"

There's that stigma again. I was infected too.

Clauser's experiment had shot down the hidden variable idea ... nearly. Quantum theory was vindicated ... nearly. Unfortunately, however, there was a loophole—a loophole through which the hidden variable concept might just possibly manage to squeeze.

An analogy to Clauser's experiment is a variation of my angry couple, intent on disagreeing with one another. To make it vivid, imagine that they live in Kansas. One day, furious and irritable, they separate. The wife heads off to Oregon. When she arrives she encounters an individual (named Alice) who for some reason is full of questions. The questions keep changing. "Do you like steak?" Alice might ask. Or alternatively "do you like fish?" or "do you like exercise?" Meanwhile, the husband has just arrived in Florida, where he is bombarded with questions by Bob—questions that are sometimes the same, but sometimes different, than those asked by Alice. "Do you like steak?" might well be the first of Bob's questions—but it also might be "do you like fish?"

As before, the husband and wife are intent on disagreeing with each other (not always now, but by a certain definite amount). The problem is that they don't know how to do it. After all, neither one of them knows the reply the other has given. They don't even know what question the other has answered! So how can they synchronize their replies?

Here's a way. They can phone one another.

That is the loophole. If husband and wife could tell each other what the questions had been, and what their replies had been, they could synchronize those replies. Some sort of "telephone connection" between them would accomplish this. There is nothing in all of physics that explains just how this connection might work. Of course it's not a matter of actual phone calls from one quantum particle to another: there's no such thing. It would have to be something else: something that has never been thought of before. But so what? Maybe it's possible after all.

That loophole is a vulnerability in Clauser's experiment. So his conclusion was open to attack. Perhaps quantum mechanics was not the right theory after all. Perhaps hidden variables actually did exist.

But several years after Clauser's experiment, a French physicist named Alain Aspect found a way to close that loophole. He did this by blocking the phone calls. He rendered those telephones—if they existed at all—irrelevant.

Remarkably Aspect managed to do this even though he knew very little about how those hypothetical phones might possibly work. Of course, that's a hard job. Normally, in order to defend against an attack, you had better know something about the nature of that attack. How can you defend against an unknown enemy?

Aspect took advantage of the fact that he did know one little thing about that enemy: it could not travel faster than light. The signals from one quantum particle to the other, whatever they might possibly be, had to obey that cosmic speed limit.

**Figure 12.2**
Alain Aspect. Clauser's experiment had a potential loophole: that somehow the two entangled particles could communicate with one another. Aspect closed that loophole by randomly changing the "questions" asked of them.

The principle that no signal can travel faster than light is enshrined in physics. It has been experimentally tested over and over again, and always found valid. Not even the weird quantum world can violate it. Aspect found a way to use this principle in his experiment. He created a situation in which the husband and wife would set forth on their journeys, one to Oregon and one to Florida—perhaps carrying telephones of some unknown design, and perhaps talking with one another as they traveled. But because those phone signals were not traveling with infinite velocity—the speed of light is great, but not infinite—there would be a tiny interval of time the transmissions took to travel between husband and wife. And in that tiny interval, Alice and Bob would *change their questions.*

In such a situation, the husband and wife would find themselves forced to answer a question before they could exchange information. Even had they possessed telephones, the transmissions would arrive too late. As a consequence, each reply would be given in a state of total ignorance. Aspect would have rendered those hypothetical telephones irrelevant.

He built the experiment. Then he ran it. He found that the new twist made no difference. His data showed that, astonishingly, husband and wife still managed to synchronize their responses. They disagreed more often than could be accounted for. They persisted in doing the impossible.

I will say it again: they were doing something for which there is no possible explanation.

So Aspect closed the loophole. Unfortunately, however, there is not just one loophole. There are many.

Here's another—and this is one that was closed by one of the prettiest experiments I have seen in years. It is known as the "freedom of choice loophole" and it proposes that Alice and Bob may *think* they have free will ... but actually they don't.

What does freedom of choice have to do with hidden variables? In chapter 9's discussion of a hidden-variable theory, I showed that if Bob rotates his analyzer, a certain fraction of the detections will show disagreements with Alice's result. But quantum theory predicts more disagreements than that, and Clauser's experiment confirmed the quantum prediction. In chapter 9, I tried to alter the hidden-variable theory by having the source avoid emitting particles in a certain direction (into the wedge of figure 9.6) in order to mimic quantum theory. But, as I wrote, this attempt at a fix would not work since Alice and Bob have freedom of choice. They are free to act in any way they wish. They might elect to turn their analyzers not to the right but to the left—or not by this angle but by that. Since there is no way to adjust the source in advance to deal with every possible choice, the conclusion was that the hidden-variable idea is not going to work.

But maybe that conclusion is not so certain. For suppose that Alice and Bob are not really free to turn their analyzers in just any old which way.

Suppose their much-vaunted free will is actually an illusion. Suppose that yesterday they had been hypnotized, and today they are under the sway of a posthypnotic suggestion forcing them to swing their analyzers only in certain ways ... and suppose that the source knows about these ways.

I'm speaking metaphorically, of course. In real experiments sources don't "know" anything. And the analyzers' orientations are not chosen by people: they are chosen by machines, components of the experimental apparatus. Experimenters try to make sure that these machines behave randomly. But what if they are not fully successful? What if their so-called "random machines" are not really random? What if those machines are actually being controlled by some process of which the experimenters are entirely unaware—a process that connects both the source and the analyzers, and that deceives us into believing in quantum mechanics?

Just like Aspect's "phone calls," there is nothing in all of physics that tells us which this controlling process might be. But, again like Aspect's situation—so what? Maybe it is possible after all. That would be a loophole too.

The freedom of choice experiment did not entirely close this loophole. But it did restrict it—dramatically. It showed that this hypothetical controlling influence must not operate in the here and now. Rather it operated centuries ago, and it came from a location thousands of trillions of miles away. The experiment grabbed hold of that control, and it shoved it far off into the depths of time and cosmos.

Anton Zeilinger, a burly, affable man with an infectious sense of humor and a love of life, is a worldwide leader in work on quantum entanglement. Throughout his career he has conducted numerous groundbreaking experiments probing the many astonishments of quantum mechanics. In Zeilinger's lab in downtown Vienna, a source emitted entangled pairs of particles. (Like Aspect's, the actual experiment worked with photons instead of electrons, and it measured their polarizations instead of spins.) A third of a mile away was a bank. We can call it "Alice's bank" if we wish. One evening a group of physicists invaded that bank. But they were not there to steal. Rather, they were there to assemble two sets of scientific equipment, peering out of two different windows.

One of those windows had a good view of Zeilinger's lab. Through the window peered a device capable of revealing the polarization of an incoming photon, a photon shot out from his lab. Out a different window, one facing

**Figure 12.3**
Anton Zeilinger. Clauser's experiment had another potential loophole: that the "questions" asked of the two particles only seemed random, but were in fact being dictated by some unknown mechanism. Zeilinger's experiment showed that this mechanism, if it existed at all, lay far off in the universe and operated far back in the past. Photo courtesy of the Mind & Life Institute, © The Mind & Life Institute.

in the opposite direction, peered a telescope. It was peering, not at a lab, not even at any earthbound building, but up into the sky. It was gazing at a star.

That telescope was not one of those mighty instruments so beloved of astronomers, perched on mountaintops or orbiting the earth, but rather the sort of small, unassuming device that an amateur astronomer might own. Indeed, the telescope was not the hard part of the experiment. The hard part was what it was connected to—and this was the sort of stuff no amateur could afford. Part of that stuff was an instrument that observed individual photons of the light from the star. At any instant a star—or any other source of light—is emitting a vast flood of photons, all of different colors. The experiment's equipment was set to grab those photons one by one ... and measure the color of each. Was it more nearly red, or more nearly blue?

Information about that color was routed across the bank to the analyzer trained on Zeilinger's lab. And there it entered the most extraordinary of devices: a device that set the orientation of that analyzer according to the information from the telescope—and that was capable of changing its orientation in a millionth of a second. And that was the key element of the experiment: *the orientation of the analyzer was set by the color of the starlight.*

All this equipment peering out of the windows in the bank constituted our "Alice." As for "Bob," he was located in a different building—a mile away, on the far side of Zeilinger's lab and its source of entangled photons. In Bob's building stood similar equipment, with the analyzer catching the second member of the entangled pair, and with the telescope pointing to a second star, whose photon would determine the orientation of Bob's analyzer.

In this way, the group had devised a setup in which the choice of orientations of the analyzers—the questions to ask of the husband and wife—was determined not by the choice of the experimenters, not by the action of some piece of equipment situated in the lab, but rather by infinitesimal bits of light from two different stars as they twinkled in the evening sky over Vienna.

They ran the experiment. It got results in agreement with quantum theory and opposed to the hidden-variable theory.

If we are talking about a loophole involving free will, the "will" we are talking about is that of those two stars. It was they that were directing the experiment, directing by means of photons launched centuries ago. One of the stars was 600 light years away, which amounts to thousands of trillions of miles. And the light it emitted had been sent forth on its journey toward Vienna 600 years in the past. The other star was more distant still.

Could our hypothetical "preordaining influence"—our hypnotist—have intervened in the experiment to invalidate its results? It could have done so only by controlling those bits of starlight. That is to say, only by intervening not in Vienna, but far off in the Milky Way. And it was not even intervening now: it had done so centuries ago ... at a time, in the words of one of those experimenters, "back when Joan of Arc's friends still called her Joanie."[17]

And just like Aspect with his experiment, the freedom of choice group found that their new twist made no difference. Their data showed that, astonishingly, husband and wife still managed to synchronize their responses. They disagreed more often than could be accounted for. They persisted in doing the impossible.

Not too many years ago a graduate student had a wonderful idea. He decided to invent a game. A metaphysical game.

The student's name was Carlos Abellán. He worked in a research group led by Morgan Mitchell, based in Barcelona. For years the two of them had been batting around the whole idea of randomness.

**Figure 12.4**
Carlos Abellán (left) and Morgan Mitchell (right). Photo: ICFO.

Are the "questions" asked of the two particles really random? All previous experiments had relied on some physical mechanism to achieve randomness—but mechanisms obey the laws of classical physics, and so are not truly random. In the "the BIG Bell Test experiment" vast numbers of people were enlisted to use their free will to create randomness.

**Figure 12.5**
The app they created. Image: Maria Pascual (Kaitos Games).

Randomness is a key element of any experiment aiming to test Bell's Theorem. It is the only way to ensure that our angry couple, intent on disagreeing with one another, have no way of knowing the questions they are about to be asked. And up to that point all the various Bell-test experiments had achieved this randomness by mechanical means. They used marvelous and elaborate mechanisms that were designed to behave unpredictably as they dictated the orientation of the analyzers. Even the distant stars in the experiment I have just described were at heart mechanisms—the fact they were natural rather than artificial was irrelevant. But Abellán and Mitchell found themselves wondering: are *any* mechanisms truly random? Or do they only seem to be?

My smartphone tells me that apps exist that behave randomly. Indeed, I can buy chance. Just now I checked the App Store, and there I found all sorts of random-number generators. I could download any one of them: each time I asked, it would give me some unpredictable number.

But are these numbers really unpredictable? No, they are not. We tend to forget that all the marvelous stuff we find on the Web rests in the last analysis on actual, physical machines. We say that those random-number generators reside "in cyberspace"—but cyberspace is not a real thing. It is a term we use for radio signals traveling this way and that through an elaborate network connecting our smartphones with servers—and each of these servers is a computer, an actual, physical device composed of actual physical parts. Somewhere, deep in the guts of a rack of electronics in some faraway server farm, tiny electrical currents flow this way instead of that, and tiny magnets are orientated one way instead of another ... and the working of this immense collection of electromagnetic parts is the working of cyberspace. If I knew exactly the physical configuration of that server I would find that my vaunted random-number generator was only an apparently random-number generator.

Maybe this will become more evident if we consider the act of flipping a coin. It is the quintessential example of randomness. Can I predict how that coin will land? Of course not. But is it random? No, it is not.

Suppose I knew exactly how high I had tossed that coin. Then I would be able to predict how long it would take before landing. And suppose I knew exactly how much spin my thumb had imparted to it. Then I would know how rapidly it was rotating during its flight, and how many times it had spun over in that interval of time. And if I knew how hard it landed, and at what angle it had struck the table when it did so, then I could predict how high it would bounce and how many times it would flip over before finally coming to rest. And if I knew whether that coin was showing heads or tails just before I flipped it ... why then, if I knew all these things, I would have been able to predict what that flipped coin would show when it landed. And make a million dollars.

For in truth, a flipped coin does not exhibit randomness. Neither does my so-called random-number generator. What they exhibit is complexity.

You might be objecting that my "research projects" into the flight of the coin or the workings of my app are not something that I could carry out in practice. I agree—but so what? We are not speaking of "in practice." We are speaking of "in principle," and the principle is one of absolute determinism: everything that happens in the large-scale world is dictated by the inflexible law of cause and effect. And if it is dictated then it is in principle foreseeable ... and in that case what our mythical angry couple is doing may not be so very mysterious after all.

That's another loophole.

Abellán and Mitchell wanted to nail that loophole shut. They wanted to achieve something no machine could do, and achieve true randomness. They asked themselves: what were the most erratic things in the universe?

People were, they decided.

You and me. The butcher and the baker and the candlestick maker and everybody else too. All of humanity, in the messiness and unpredictability of free will. Never mind those coin-flips and servers and distant stars: Abellán and Mitchell would assemble a team to build an experiment in which it was not a physical mechanism that chose how to rotate the analyzers to and fro. It would be people—ordinary people, people from every walk of life—who made those random choices.

It was not a new idea. Many researchers had already bandied about the notion of replacing machines with humans. But nobody could figure out how to make it work. The problem was that people were not fast enough. They were capable of making choices only so often—a few times per second, maybe. But the experiments needed to flip their analyzers' orientations at lightening speeds.

Abellán's wonderful idea was to circumvent this limitation by using large numbers of people. How to get in touch with them? Use social media. And how to persuade them to make those random choices? Lure them in. Invent something so attractive, so seductive and enticing, that vast numbers of people would be sucked into the project. How to do this? By inventing a game.

It would be an online game. The researchers would create a network of players, a vast agglomeration encompassing huge numbers of people from across the globe, all playing the game at the very same time and each one of them making random choices. Out of this network of players an immense storehouse of pure chance would accumulate. Accumulate, and drive the course of experiments testing Bell's Theorem.

They called their game "the BIG Bell Test." It lived on a website.[a]

The test would ask people to make choices. It was actually very simple, if truth be told: all a player had to do was enter a series of 0's and 1's into a smartphone. The hard part was that this had to be done randomly and rapidly.

a. You can play this game yourself. See https://museum.thebigbelltest.org/#/home?l=EN

The experimenters built into their app all the elements of modern gaming: trendy animations and sound effects to cheer the gamers on their way, leaderboards, boss battles and power-ups, the opportunity to form groups and compare their skill levels with those of fellow-gamers. From time to time players would be reminded that their inputs were being used in actual experiments underway at that very moment in laboratories across the globe. And as they refined their skills and graduated from one level of the game to the next, they might be rewarded with some interesting tidbit about the mathematics of randomness, or by a prerecorded video from one of the experiments.

Meanwhile the Oracle would be watching.

The Oracle's function was to tell the players how well they were doing—how much randomness they were achieving. For it turns out that it is not so easy to be random. Suppose for instance that, as you madly typed away, you just happened to enter three 0's in a row. Studies have shown that in such a situation you are more likely to avoid 0 for your fourth step and enter a 1 instead. This in spite of the fact that true randomness dictates that you should be equally likely to choose either. The Oracle was a "prediction engine" that studied your previous entries and tried to anticipate your next: if it succeeded, this was proof that you were not achieving true randomness. And the Oracle would tell you so. The Oracle was the enemy against which you were competing.

Having built their game, the next step was to recruit the players. (They called them "Bellsters.") A massive advertising campaign was launched—on social media, in newspapers and TV ads and announcements to schools and science museums. More than 230 headlines resulted. The game and accompanying information were made available on the group's website in seven languages, making it accessible to over three billion people: that is nearly half the world's population.

Of course, all this was only the first step. What about the experiments that would use the data? The members of Mitchell's group were planning to run their own experiment. But they wanted others. They advertised their project to laboratories across the world, and ultimately assembled a group of researchers running fully 13 different experiments. They were situated in nations spanning the globe: Barcelona, China, the United States ...

Everything would happen at the same time: experiments running, Bellsters gaming. The whole thing was a massive exercise in organization. Gaming day was set for November 30, 2016.

Across the spinning globe daylight dawned. The experiments fired up and got ready for the data. Meanwhile, the gamers got busy. They came from most of the nations of the world: from Europe and the Americas, from China, Australia—there were even some from Antarctica. Over the course of the next 51 hours some 100,000 gamers entered their data: more than 97 million 0's and 1's. Over one 12-hour period the worldwide group reached and sustained a rate of 1,000 random choices per second.

The 13 experiments, all running at the same time, used these data to test Bell's Theorem. Every one of them found that quantum mechanics was valid and hidden variables were not.

It all rests on a conundrum, of course. For is it really true that we have free will? Were the gamers actually behaving randomly? I think it is fair to say that our brains are machines. and so cannot be truly random. But a *brain*—the actual, physical object lying between our ears—is one thing, and a *mind* is another. Is it possible for our brains to be deterministic while at the same time we humans are not? As for myself, I have no idea.

So we can make of this project what we will. Perhaps it adds a vital element to the situation and perhaps not. But no matter: it was a wonderful project—wonderful for the gamers and, I am willing to bet, for the scientists too.

Nowadays many groups of experimentalists are hard at work, closing loophole after loophole. There are certainly enough of them to be closed. Why only last week I came across an article listing fully ten. The task of closing all these loopholes is not yet complete. Indeed, it is not even a matter of closing first one and then another: best of all would be to close them all at once. So it's a slow process.

But perhaps you feel that all this effort is just a little bit silly. After all—aren't these physicists being maybe just a little bit paranoid? Are they perhaps starting to resemble some bunch of conspiracy theorists, hard at work inventing one nutty idea after another? Who could possibly take seriously the notions of "phone calls between particles" or "preordaining influences"—notions without the slightest thing to recommend them beyond my lame assertion that "maybe it is possible after all"?

I'll tell you who would do all this: anybody who wants to be sure about quantum theory—very sure, as sure as humanly possible about one of the most important scientific discoveries of all time.

Scientists want their knowledge to be trustworthy. All of us want our knowledge to be trustworthy. So much of life is uncertain. We want as much certainty as we can get. We want to be able to trust what little we do know.

I like to think in terms of the analogy of climbing a ladder. Suppose you have propped a ladder up against a wall—a long ladder, one that will carry you way up into the heights. Would you be willing to start up that ladder before making sure it is secure? As for me, I certainly wouldn't. I would shake it, swing it to and fro.

In doing so, my goal is not to knock the ladder down. It is to make sure that nothing else can knock it down. That's what these loophole-chasers are doing. They are "shaking" quantum mechanics in every way they can. They are probing for weaknesses in all the various experiments that purport to confirm the theory—all in order to make sure that the experiments are trustworthy. And the results to date have shown that the ladder that is quantum mechanics is utterly trustworthy.

But I would not want to leave it at that. For in truth I believe that there is a further reason these people are spending so much time and effort on all these beautiful experiments. It is that the experiments really are beautiful. It is that only now, now that the latest and sexiest gadget is available from that high-tech corporation in Texas; and only now that those colleagues down the hall have invented yet another brilliant technique for doing yet another new thing—only now has it suddenly become possible to do what yesterday was impossible. So you get down to cases and you do do it.

I love that about science: that great, windblown sense of openness about the whole enterprise.

# 13 ... Metaphysics

Clauser was not happy with the result of his experiment. He had been thinking that his gadget would yield the opposite conclusion. Indeed, he once told me that he had been so sure of things that he was willing to place a bet that quantum mechanics would turn out to be wrong. Two-to-one odds were the best that a colleague would give him, and he accepted them. Unfortunately—or fortunately, if you prefer—Clauser lost the bet. He mailed off a two-dollar bill to his colleague: so far as he knows, that colleague still has it up on his office wall.

After all, Clauser had wanted to discredit quantum theory.

> I was convinced that quantum mechanics had to be wrong.... I kept saying, "Well, we did the experiment, what could be wrong?" Obviously we got the "wrong" result. I had no choice but to report what we saw—You know, here's the result. But it contradicts what I believed in my gut has to be true. The result, I didn't expect. I hoped we would overthrow quantum mechanics.[1]

But it was not just a matter of youthful rebelliousness. It was not just a matter of wanting to overthrow a cherished theory. It was also a matter of having to figure out what his experiment was telling us.

Quantum mechanics predicts the impossible—we have known that for decades. But what Clauser's and Aspect's and all the other metaphysical experiments are telling us that *the real world* accomplishes the impossible. And how can that be? What have we learned from all the experiments testing Bell's Theorem? Nature violates Bell's restriction: what does this astonishing result tell us?

The question is hard to answer in any simple way, and there is no agreement among workers in the field. We are still sorting it all out.

One possible conclusion to be drawn is that there are no hidden variables. There is no real physical situation, no actual state of affairs. If this is so, it means that my Great Predictor is silent for a reason: that there is simply nothing to be said more than what he does say. It also means that quantum mechanics is not a half-theory at all, but a full theory, a theory perfectly suited to the strange new world on which we have inadvertently stumbled.

This conclusion is not airtight, for it rests on an assumption—the so-called "locality" assumption—which I will discuss later. But for now, let us ask what it could possibly mean.

It means that particles in the quantum realm do not possess certain properties. But I would argue that properties are essential for our thinking. They are built into the very way our minds work. How could we think of an electron spinning but not spinning in a definite direction? How might we imagine a particle with not zero speed, not with this speed or that speed, but with no particular speed ... but nevertheless produced at a certain moment at a certain place, and detected a certain amount of time later a certain distance away? How could we hold in our mind's eye the image of a thing without a location? It all seems like a very violation of logic.

It is not like forgetting where you parked the car. You think that the car might be on this street or that street—but imbedded in this way of thinking is the conviction that the car is somewhere, at a place that exists but that you do not happen to recall. This is different. This would be a car without the property of location. Without location until you find the car, at which point its whereabouts become entirely real.

To say that hidden variables do not exist is to call into question the very meaning of what we mean by a *thing*. For surely, things have properties, and these properties have consequences. Consider:

- A living room has an open window. Stand inside the room, and toss a ball in some random direction. Does the ball make it through the window, or clatter against the wall and remain inside? The property that determines the outcome is the direction the ball is moving. That property is a variable: if the lights are out and you can't see the ball, it is a hidden variable.
- A ray of light is approaching a piece of red transparent glass. If the light is red it gets through the glass: if it isn't, it is blocked. The property that determines whether the ray passes through the glass is its color.
- A woman is approaching a bar. On the door there is a sign: "No one under 18 admitted." The property determining whether she gets in is her age.

- In chapter 2 I gave an analogy to radioactive decay: the analogy of a maple tree in autumn. Some of the leaves fall sooner than others. But why? When I questioned the Great Predictor, asking for the reason, he would not answer. And now we know why he so adamantly refused to speak: because there was no reason. A "reason," after all, is a hidden variable ... and hidden variables do not exist.
- An electron is approaching a detector oriented along the vertical direction. Will the detector find "up" or "down?" Here apparently there is *no* property belonging to the electron that determines what the detector does. But nevertheless, the detector does something.

In many ways, electrons seem quite ordinary. An electron can be produced—by an electron gun, say—at a certain place and time. It can be detected—by the screen on a TV set, say—at another place and time. An electron has a perfectly definite mass and electric charge and magnitude of spin.

All this makes us think of an electron as being a *thing* in the ordinary sense of the term: sort of a tiny pebble. But nobody has ever seen an electron—their presence is only inferred indirectly—and maybe we are a little hasty in treating them so cavalierly. For consider *good news*. *Good news* can be produced at a certain time and place, it can be detected by a person at some other time and place, it has an effect on that person, and it travels quite rapidly. Nevertheless, nobody would think of it as a thing. Maybe an electron is more like news, and not so much like a pebble.

But it is only *certain* properties that the electron does not possess. The particle has a perfectly definite mass and charge, for instance. Furthermore, it is not only electrons that we are speaking of here: photons, neutrons, atoms ... *every* denizen of the microworld partakes of the same enigmatic quantum nature.

There are times when I think that what we really need is a new terminology. We speak of particles as *things*. We say that some*thing* left the electron gun and arrived at the detector. But when we speak in these terms, we are naturally led to ask all sorts of questions about these particles. Why can't we see them, for instance? And what shape are they—spheres, cubes, or perhaps shaped like some Chinese ideogram? What is their color? Are they cheerful or gloomy, sweet smelling or acrid? Carrying on in this way, we can be easily led astray—led astray by the set of unconscious associations that the word "thing" raises within our minds. Our language forces on us a certain way of thinking, a way that apparently we must be careful to resist.

Every time a quantum particle enters a detector, the detector responds by doing something. Maybe it tells us that the particle's spin is along some direction. Maybe it tells us that the particle is over there. Maybe it tells us that the particle is zipping along at such-and-such a velocity. But what do these responses mean? A measurement is supposed to reveal a property of the thing studied—a thermometer tells us the temperature of the air, a barometer its pressure. These are not matters of opinion, not matters of taste, but facts: real properties of a real world. But if there is anything we have learned so far, it is that the microworld is different. We used to think that a particle has a spin that points in some direction, that it is at some place, and that it is moving with some speed. And we used to think that the detector has merely found out these properties. But now ... well, now we had better be careful.

Because if things do not have these properties, then what has a measurement told us?

Quantum theory has an answer to this question. The answer is that a measurement does not *reveal* a property of the microworld: rather, the measurement *creates* that property. Prior to the measurement the electron spin had no particular direction: after the measurement it does.

That is a gigantic shift in thinking. Do you like that shift? Is it congenial to you? Read what the brilliant physicist E. T. Jaynes has to say about it:

> From this, it is pretty clear why present quantum theory not only does not use—it does not even dare to mention—the notion of a "real physical situation." Defenders of the theory say that this notion is philosophically naïve, a throwback to outmoded ways of thinking, and that recognition of this constitutes deep new wisdom about the nature of human knowledge. I say that it constitutes a violent irrationality, that somewhere in this theory the distinction between reality and our knowledge of reality has become lost, and the result has more the character of medieval necromancy than of science.[2]

The ideal of science is that we are investigating a real situation that exists independently of us. But if we are not ... then what *are* we scientists doing?

# 14 Nonlocality

What are we scientists doing? Before quantum mechanics came along we would have replied that we are studying the properties of the world. But if there is anything that we have learned from Bell's Theorem and the experiments that test it, it is that the microworld does not necessarily have certain properties.

To be specific, let us return to the EPR scenario in which Alice and Bob's detectors are parallel. In this configuration they always get opposite results. But why? We used to think that it was because the two particles heading toward those detectors had spins pointing in opposite directions. But as we saw in chapter 9 that simple picture does not work. Furthermore, as Bell's Theorem and the experiments that test it have shown, if we make the locality assumption *no* picture will work that attributes definite properties to those particles. So once again—what makes Bob's detector get different results than Alice's?

It used to be a trivial question. Suddenly it is not so trivial.

If the answer to that question does not involve the properties of the particle heading toward the detector, then it must involve something else. What else? There is only one possibility. This possibility has to do, not with particles, but with measurements. It is that in some strange way Bob's result is connected to a result—the result of Alice's measurement. We are forced to conclude that *the very fact that Alice's detector gets one result influences Bob's to get the other.*

We must cease thinking about the particles heading toward detectors, and start thinking about something else—about the behavior of these detectors. We must realize that these behaviors are connected—invariably connected, perfectly connected. Our discovery is that Alice's and Bob's detectors always behave in ways that are synchronized, and that they do so even if there are no wires leading from one to the other, even if there are no radio transmissions from one to the other, and even if they are thousands of miles apart. We must understand that the world is utterly connected.

Physicists term this connection "nonlocality." Things happening far away are linked to things happening right here.

In proving his result, Bell was careful to analyze theories in which nonlocality had no place. It is only local theories that his theorem and the experiments that test it have ruled out. If, on the other hand, we recognize that the world actually is nonlocal, then Bell's Theorem has no validity. In such a case, particles in the quantum world can possess perfectly definite properties.

(A theory along these lines was long ago developed by the physicist David Bohm—indeed, it was by thinking about this theory that Bell was led to his discovery. Within Bohm's picture quantum particles have perfectly definite attributes. But even so, his world is utterly unlike the normal world of daily experience, for it is profoundly nonlocal.)

Quite aside from Bohm's theory, nonlocality denotes an intimate connection between widely separated events. At first glance this linkage might not seem so very strange. Perhaps it reminds us of the utterly connected nature of everyday life, in which we stay in touch with friends through Facebook, follow events in China through CNN, and buy avocados grown in Mexico. But quantum nonlocality is not like all this. It is not like anything we have ever encountered before.

On the one hand, the nonlocal influence must be able to exert itself across gigantic gulfs of space. This is because Alice and Bob get opposite results even if they are very far away from one another. Nonlocal connections grow no weaker with distance. Even were Alice located on some distant planet in a faraway galaxy, this invisible agency must be able to exert its controlling sway.

Furthermore, it must do so instantaneously. Our daily connections, whether by telephone, internet, or the like, travel at the speed of light or slower—but this influence must travel faster than light. For suppose the two electrons in our experiment were set forth on their journeys from a point half-way between Alice and Bob—and then Alice were to take one small step forward. She would receive her electron a fraction of a second before Bob. So the influence we are postulating must travel from her to him in that fraction of a second. Indeed, Alice and Bob could be located at enormous distances from one another. Alice's home might lie in a galaxy a million light years distant, so that a ray of light from her to Bob would require a million years to arrive—but her influence would still get there in no time flat.

Indeed, we are forced to postulate that our mysterious influence travels at a literally infinite velocity. So this strange new phenomenon has nothing to do with the "telephone calls" between particles that Alain Aspect's experiment had dealt with (chapter 11). It is another matter altogether.

And yet, according to Einstein's theory of relativity the very concept of "no time flat" has no meaning—because while two events may happen at the same instant to one observer, they do not for another. Many people believe that we are facing here a major conflict between the two great discoveries of twentieth century physics, relativity and quantum theory.

And finally, it is not at all clear who is doing the influencing. If Alice receives her particle first we might be willing to say that the result of her measurement caused the result of Bob's. But if she takes a few steps back then Bob would be the first to register a measurement. Is it now Bob's detector that is calling the shots? And finally, what would we say if Alice and Bob receive their particles at the very same instant? Then what is influencing what?

The lesson we must take from this is that we cannot think of one result "causing" the other. We must think only of a synchronization between the results. Of a correlation between the behaviors of the detectors.

Doesn't this correlation violate Einstein's principle that nothing can travel faster than light? After all—something that travels at an infinite velocity certainly *seems* to be achieving this remarkable feat. It does not—for three reasons. In the first case, Einstein's principle applies to objects (spaceships and the like) but our postulated influence is not an object. In the second case, Einstein's principle applies to causes—to physical processes that exert a causative effect. But the influence we are postulating is not a cause in anything like the ordinary sense of the term. There is nothing Alice can do to make Bob's detector do anything. She cannot cause his detector to obtain a certain result—because she cannot cause *her* detector to obtain *its* result. It is not Alice who influences Bob's detector: it is the result she obtained. And finally, Einstein's principle applies to information—to messages that we send one another. But even though Alice and Bob might have a prior agreement that, say, receipt of an electron with spin up means "sell all your stock in Facebook," the fact that Alice cannot control the result Bob's detector gets means that she cannot control the message he gets … and an uncontrollable message is no message at all.

Perhaps strangest of all is that Alice and Bob might not even know that their electrons are connected in this strange fashion. If Alice studies only her particle, nothing she can do will alert her to the fact that it is associated with Bob's. The same applies to Bob. Both experimenters believe themselves to be studying isolated, individual particles. Only if they were to get together and compare notes would they realize that the electrons they were studying were actually connected. The same applies to any pair of particles. Perhaps the electrons in your prefrontal cortex are intimately linked with those in the brain of that person across the room—a person you have never met before. Or perhaps they are linked with electrons in the body of some alien creature living on a world in a distant galaxy of which we are entirely unaware.

The message of nonlocality is that the world is utterly connected. The fall of a tree in Chile might be linked with the rising of a plume of dust on Mars. The fall of a sparrow in Norway might be linked with the birth of a baby next door.

# 15 Quantum Machines

The birth of a baby next door ... or the secure transfer of funds from Vienna's city hall to the Bank Austria Creditanstalt—a transfer initiated by the city's mayor, executed by the bank's director, and announced at a press conference. That was in 2004.

What began as a philosophical difference among a small group of physicists nearly a century ago has blossomed into what promises to be a worldwide industry—an industry based on quantum mechanics. Billions of dollars are involved. Google is interested in quantum nonlocality. So are Facebook and MIT. Venture capital firms are sitting down at the table, as well as the CIA. Above the boardrooms of mighty governments and corporations float the ghosts of Einstein, Bohr, and Bell. Mild, philosophical, perhaps a bit otherworldly, they have been shoved aside by the new breed: those can-do types who roll up their sleeves, brush aside the niceties, and get down to cases. The old stigma has passed—passed with a vengeance.

But why? What has happened to the old antipathy to philosophically tinged questions, the antipathy that stifled the field for years? Part of the answer is a simple matter of time. A new generation of researchers has come of age—researchers no longer in thrall to Von Neumann's erroneous proof or to the necessities of the Cold War. These people never experienced the old days, were hardly touched by that subtle, all-pervasive amusement that once greeted those who asked such questions—an amusement I must have unconsciously sensed when I was a student, and that steered me away from these matters for so many years.

But I think there is another factor at play. As I wrote in chapter 12, I think it is also a matter of the advance of technology. In the intervening years technology has progressed so rapidly that it is now possible to actually do the experiments of which previous generations could only dream. Thought experiments have been replaced by real ones. It is no longer a matter of arguing about what such-and-such an experiment might reveal: now you can actually find out. There it is again: the thing I love about science—the wonderful sense of freedom and openness and possibility to the business of research. If you *can* do something you *do* do it, and all those psychological and historical issues be damned. Science is a can-do enterprise. So philosophy is invading industry.

And strange to say, all this has been made possible by a breed pretty much uninterested in philosophy. The new age of experimental metaphysics has been made possible by people in love, not with philosophy, but with gadgets. These are people who delight in inventing new devices, new procedures, and new ways of doing experiments. If their wonderful new gadgets can be pressed into service to answer a primarily philosophical question, all well and good. But it was never their primary concern. So industry is invading philosophy.

And so experimental metaphysics.

If you are transferring funds you are transferring information—information about your bank account number, let us say. This might be done in person, or by email. But if your name happens to be Alice, and the teller at the bank is named Bob, a whole new dimension of the situation just might occur to you. Can information be conveyed by quantum particles?

When we measure the spin of a particle, we learn whether that spin is along or against the direction of our detector. Suppose we agree that a spin along represents a "0" and a spin against a "1." Then our measurement has told us a number.

That number is in binary—the number system of base two. We are used to writing numbers in base ten. But the translation is straightforward:

| Base 10 | Base 2 |
|---|---|
| 0 | 0 |
| 1 | 1 |
| 2 | 10 |
| 3 | 11 |
| 4 | 100 |
| . | . |
| . | . |
| . | . |

Suppose Alice sends Bob three electrons, the first with spin against his detector's axis–that's a "1"—and the next two along the axis—these are "0s." Then this represents the sequence "100"—which, if we think of it as the binary representation of a number, we would interpret as a "4." Using those electrons, Alice has told Bob a number.

We can also transfer letters of the alphabet. It can be so simple a matter as agreeing that "1" represents the letter "a," "2" represents a "b," and so forth. There are more sophisticated codings, but the principle is the same. In every case we can find a means of translating the information we wish to convey into binary numbers, and then we can use electron spins to encode those numbers. Alternatively, we can use particles of light—photons.

It's not enough to transfer information. The transfer needs to be secure. We need to make our information available to the intended recipient, but not available to anybody else. The world is full of eavesdroppers—hackers trying to steal our credit card numbers, wartime enemies trying to steal our battle plans. If you are a certain soft drink company, you might wish to keep the formula for Coca-Cola secret from competitors, while revealing it to your factories worldwide. How to guard against snoops?

Secrecy in the transfer of information has a long and fascinating history. In one ancient method, the head of a courier was shaved, the message was written on his scalp, and his hair was allowed to grow back. The courier could then travel to the intended recipient, where his head was shaved, thus revealing the message. A more recent technique was to lock the secret message into a briefcase, handcuff the briefcase to a courier, and send the courier off to the recipient—a recipient who had the only key to the briefcase. I recall sitting next to such a courier once in an airplane.

Both of these methods entailed trusting the couriers—and a long trip too, one that might very well be expensive and time-consuming. Far cheaper, and far more rapid, would be something like a telephone call or a transfer over the Web. How to guard against eavesdroppers in these situations?

The method is to encrypt the message—to scramble it in some way. Suppose we send the message "MN." An eavesdropper could easily intercept it, but she would not have the slightest idea what it means. If, however, you and your intended recipient had agreed beforehand that you would send the letter of the alphabet lying *before* each letter of your message, the recipient would know that in fact you had sent the word "NO."

Such a method of encryption is known as the "key"—it unlocks the message. This one, of course, is very simple, and it is one that any smart eavesdropper could foil with ease. It is far better to use a more complicated key. Suppose, for example, that we use some random string of numbers, such as

726 ...

and suppose we simply add each digit to the corresponding digit in the message we wish to send. If, for instance, Alice's credit card number begins with the digits

547 ...

then she would add the key to her number

7 + 5, 2 + 4, 6 + 7 ...

and send the message

12,6,13 ...

If Bob knew the key, he could decode the message and so learn the first three digits of Alice's credit card number. And if nobody else knew the key, eavesdroppers would be foiled.

Of course, Alice has to tell Bob about the key she used—and this "telling" itself must be secret. There is a whole branch of mathematics devoted to the study of encrypting and decrypting messages. It is known as cryptography. No cryptographic method is foolproof. Some brilliant new technique is invented ... and then, far sooner than anybody expected, some brilliant hacker finds a means of circumventing it. For obvious reasons, large corporations are interested in cryptography, as are governments. Billions of dollars are involved.

People's lives are involved as well. During the Second World War, Germany encrypted its messages using a devilishly complicated device known as the "enigma" machine. In absolute secrecy, code breakers stationed at Bletchley Park in England succeeded in deciphering the code, revealing vitally important military plans. Their triumph shortened the war, probably by several years, saving untold numbers of lives. Had they failed, would an atomic bomb have been dropped on Berlin?

In 1984 two researchers, Charles Bennett and Gilles Brassard, invented a way to encrypt a key relying on quantum mechanics. Their method is to use our old familiar pair of entangled particles, for which measurements of spins always yield opposite results. Suppose we send out a series of such entangled pairs. For each one, Alice and Bob can translate the results of their spin measurements into binary numbers—and then, if Bob simply reverses his digits, his number is guaranteed to be the same as Alice's. After measuring all the particles, Alice and Bob will have the same series of numbers—the same key. If Alice uses her key to scramble the message she wishes to send, Bob can use his to unscramble it.

Of course an eavesdropper might be snooping. Let's call that eavesdropper Eve. Eve could intercept each particle heading toward Bob, measure its spin, and so learn each digit of the key. Furthermore, she could disguise her presence by producing a new particle of her own, a particle with the same spin as the one she has just measured, and sending it off to Bob. Alice and Bob would be unaware that their supposedly secret key had been intercepted. But in a brilliant invention, Bennett and Brassard figured out how to get around this danger. Their method is to have Alice and Bob randomly twist the axes of their spin detectors about, sometimes using one orientation and sometimes another. Through a complicated series of steps, both sender and recipient would be able to learn if Eve was present. Furthermore, Eve would be prevented from knowing whether the information she received was real or bogus.

Bennett and Brassard did not just invent an idea—they teamed up with a colleague and actually did the trick. Their device (it happened to use particles of light) transferred information along a 30-centimeter box, cheerfully dubbed "Aunt Martha's Coffin." That was in 1984. Over the next few decades the field matured. By 2005 four companies had been formed to develop commercial products: prices ran into the hundreds of thousands of dollars. Shortly thereafter the state of Geneva announced its intention to use quantum cryptography to ensure the security of its network linking ballot data entry during the Swiss national elections.

Anton Zeilinger (we first met him in chapter 12's freedom of choice experiment) was the leader of the 2004 demonstration involving the bank transfer of funds. His bank-transfer demonstration involved a collaboration with a corporation interested in creating a marketable quantum cryptography product, a second collaboration with a corporation that laid the optical fiber along which the particles of light were sent, yet another with the receiving bank and finally one with Vienna's city hall. Within the offices of the bank pairs of photons—particles of light—were produced. One was sent down the optical fiber to city hall, while the other remained within the bank. When detectors measured the properties of the photons the key was produced and used to decode a secret message.

A milestone was reached in 2017, when a group led by the Chinese physicist Jian-Wei Pan managed to transmit a cryptographic key from interplanetary space down to the ground. The year before an entire satellite devoted to the study of quantum entanglement had been launched from China. Named after the Chinese philosopher/scientist Micius (roughly contemporary with Socrates) it flew at an altitude of some 300 miles, in an orbit designed to carry it over Beijing every night shortly after midnight.

On board that satellite was equipment designed to implement the Bennett–Brassard technique of quantum key distribution—and a small telescope, pointed downward. Below it, in a suburb of Beijing, stood a larger telescope. Executing a carefully choreographed swivel, it tracked the satellite as it zoomed overhead at 17,000 miles per hour. Within five minutes the satellite had passed over the horizon ... but in that five minutes a cryptographic key had been passed down from space.

Zeilinger and Pan are also pioneers in the field of quantum teleportation.

**Figure 15.1**
Jian-Wei Pan at an experiment. Photo courtesy of the Micius Group.

**Figure 15.2**
Launch of the Micius Satellite. Photo courtesy of the Micius Group.

**Figure 15.3**
The Micius Satellite.
Jian-Wei Pan is the leader of a group that launched a "quantum machine" into orbit about the earth. This machine, the Micius satellite, has been used to securely transmit a cryptographic key from one place to another, and to teleport a quantum state from one place to another. Photo courtesy of the Micius Group.

Every science-fiction addict is enamored of teleportation. I certainly was as a kid. I would imagine closing my eyes, trying very hard, and opening them to find myself instantly transported to Mars. I would then enthrall myself for hours imagining various adventures as I explored the Red Planet. As for how I had gotten there, and what that "trying very hard" entailed … well, this was a technical matter best left unexplored.

Now that I am older I have a clearer idea how this might be accomplished. We normally suppose that to teleport something you need it to arrive at its destination without having traversed the intervening spaces. But perhaps you do not need to actually send the object being teleported. Perhaps it might be sufficient to send nothing but information—information describing in excruciating detail exactly how your object was constructed, information sufficient to enable a factory at the destination to construct a perfect replica of your object. That might be considered teleportation.

Quantum nonlocality provides yet another a way to realize this dream—and this is a way that actually works. Zeilinger has done it. His system began with a quantum particle in some state, and it transferred that state over a distance of six city blocks. It did so beneath—*beneath*—the river Danube.

There is an island in the Danube as it runs through the city of Vienna. On that island is an underground pumping station for the city's sewage system. Once on the island you take an elevator down to that station. From it, tunnels lead beneath the river over to the mainland. Through those tunnels run the sewage pipes, together with a maze of electrical cables. One of those cables was laid by Zeilinger's group: it was a state-of-the-art optical fiber carrying quantum particles—photons.

The process producing these photons began with a laser. Housed within the underground pumping station, it was big and expensive—it cost as much as a house. The laser shone on a translucent wafer of crystal. In contrast to the laser, the crystal was fragile and tiny, a mere two millimeters thick. When illuminated by the laser, it produced a pair of photons entangled to form a nonlocal state.

One of these quantum particles stayed where it was produced. It ran through a length of optical fiber within the pumping station. The other went into a second optical fiber that led through the underwater tunnel and then up to a receiving station on the opposite bank of the river. There was another communication link involved as well: this one was entirely nonquantum, and involved nothing more than a radio signal. The signal went from a transmitter on the island to an antenna on the roof of the receiving station.

The experimenters needed their quantum particles of light to arrive at the receiving station after the radio signal. This they achieved by delaying the photons—by increasing the distance they had to travel. The experimenters did this by sending them through extra lengths of optical fiber lying coiled around on the floor.

Once the radio and quantum signals had arrived at the receiving station, delicate and complex electronics together with computer algorithms combined to yield a photon at the receiving end arranged to be in precisely the same state as the original photon that had been produced underground six blocks away. In the process, the original particle had been destroyed. But no matter: its state had been teleported. The information contained in the particle had been teleported.

Since then, the field has developed. In 2015, Zeilinger's group achieved teleportation between two islands in the Atlantic: two years later Pan's group teleported a photon from Tibet up into their Micius satellite.

Eventually teleportation might be used to connect quantum computers.

Gordon Moore, cofounder of Intel Corporation, once observed that, as they came out of the factories, the individual components of which integrated circuits were formed were growing smaller and smaller. As a consequence, the number of such components that could be crammed into an integrated circuit was increasing. Working out the numbers, he found that it was doubling every two years. So computers were doubling in power every two years.

That simple observation has come to be known as Moore's Law, and it has stayed valid. If things continue this way much longer, the components of which computers are built will have shrunk to quantum dimensions within a mere few decades. The rapid advance of technology will have brought us right into the quantum realm.

Nowadays, your computer works on "0's" and "1's: binary digits, or "bits" for short. But in this new realm the bits will be "qubits"—quantum bits. Just as electron spin can be along or against the direction of a detector, so quantum mechanics also allows it to be in a more ambiguous state—a state that cannot be described in plain language, one with no correspondence in ordinary experience, but that in some loose sense might be thought of as an electron in both configurations at once. That electron could embody a qubit that was both a "0" and a "1" at the same time. And a computer that worked with qubits would be a quantum computer.

Indeed, a qubit has not just two, but an entire range of values simultaneously, with "0" and "1" merely indicating the extreme values. Present-day computers, working only at these two extremes, do one thing at a time, or at best a few things. Their immense power stems from the fact that they do these things very rapidly. But if a qubit can be a whole range of things at once—why then, a computer working with qubits can do a whole range of things at once. That would make them faster—much faster. And this enormous increase in computing power will bring with it an enormous increase in what computers might do.

What is your favorite unsolved problem? Predicting the weather? Developing new pharmaceuticals? Mastering the vagaries of the stock market? Transferring the latest viral video to kids worldwide? You might find such hyper-powerful computers just what you need.

Governments and corporations need them too. One of the most famous methods of encoding a message involves a secret key generated by what is known as RSA encryption. That method of generating a secret key can be broken by a computer—but a current-day machine, chugging away at the problem, would take a century to crack the code. If quantum computers live up to expectations they will be able to do it in a matter of days.

So a horse race is underway. Quantum computers are in a primitive state right now, but many people are in the race, and most think their efforts will bear fruit in a matter of decades at most. Research teams at many of the world's leading universities are working on the problem.

And so are corporations. Large sums of money are involved. I recently read an article about the potential uses of quantum computers on a website devoted not to science, but to business news. The site quoted the musings of the director of engineering at Google—as well as revelations from Edward Snowden's leaked documents.

Indeed, government has always been interested in these matters. Astonishingly, as far back as 1972 the Defense Intelligence Agency developed a long and highly classified discussion of how the Soviet Union was spending great sums of money on ESP phenomena, and how the United States might want to do so as well—and, as part of this effort, to fund research on quantum nonlocality. I'm only guessing of course, but I wonder if people in

Washington were wondering whether a nuclear weapon could be rendered harmless by nonlocal influences. Or detonated.

But there is no need to wonder about one thing. Not too many years ago the United States government brought before a grand jury a researcher who had developed a powerful new method of encryption and given it away for free. They were investigating him for having disseminated a new kind of weapon.

The amazing thing about these quantum machines is how they work. For that is the point of this book, and the point of Bell's Theorem: normally explanations involve properties, processes, causes, and effects—and we now know that these concepts simply do not apply to the quantum world.

But so what? Quantum machines *do* work. They work just fine. And we know how to build them. More and more in coming years we will be using machines that live in the world of the bizarre.

# 16 A New Universe

It was many years ago that I first encountered the Great Predictor.

I was thrilled to meet him. I'd been looking forward to the encounter for years. The Predictor was famous—world famous. He was legendary for the number of his predictions, and for their amazing accuracy. Many people had relied on those predictions, and always with profit.

What intrigued me the most, however, was how bizarre were some of his predictions. "Tomorrow you will be in two places at once" was one. "On Wednesday an event will occur for which there is no cause" was another.

How could such things be? I was captivated by the strangeness of these prognostications. Could such weird things really come to pass? That's why I had been so anxious to meet the Predictor. For years I had looked forward to finally getting to know him.

At long last I was getting my wish. I was twenty years old, and I was thrilled. I was sitting in a classroom, in college, on the first day of a course called Introduction to Quantum Mechanics.

That was many long years ago. And throughout my career I have maintained my early fascination with quantum mechanics. Somehow, though, I never felt that I really understood the theory. It always sat lodged in the back of my mind—enigmatic, mysterious, enticing. Over and over again, I found myself thinking that someday I really ought to go back and figure it all out, and finally put all those early juvenile confusions to rest.

Part of that project was an effort to understand Bell's Theorem. To be honest I found myself dreading getting to work on that particular topic. While I had never felt comfortable with quantum mechanics in general, Bell's Theorem was a topic that I felt positively unnerved by. Over and over again I had tried to master it, and over and over again I had failed.

Eventually I did come to some sort of understanding of Bell's work. I recall feeling pretty pleased with things ... until the fateful day when I looked at my reflection in the mirror—and this is literally true—and I spoke aloud. "Greenstein," I said to my reflection "you were just kidding yourself, weren't you? You never really understood Bell's Theorem at all, did you?"

"Time to get going," I told my reflection.

And I did.

What did I do? I read some books. I read some scientific articles. I have already mentioned one: there were others. But, truth to say, not that many books, and not that many articles. I talked to colleagues—but not that often. I took long walks and stewed things over. Mostly that's what it was: thinking. I thought and I thought and I thought. It went on for several years.

What was I thinking about? At the beginning I could not even say. I would go back and look at the proof of Bell's Theorem. I would work through the mathematics for myself—remarkably, the actual calculations involved are not so very hard, now that Bell has shown the way. But even after I had done this I still felt mystified. The math wasn't the point—the point was what it all meant. I knew damn well that Bell's discovery was important—everybody *else* was going around saying that it was important, and it certainly felt important to me. But why? What was his theorem telling us? And why would my mind go blank whenever I tried to think about it?

That last question was a signal. By now I know myself well enough to realize that if I find it hard to even think about something, it is a signal that there is some enormous gap in my understanding. Somewhere, something was missing from my thoughts. But what?

For months it would feel that nothing was happening—but then one day I would cast my mind backward and realize that my thinking about Bell's Theorem was different. In the intervening months my thinking had changed without my even being aware of the change. As a matter of fact, that was pretty much what it was like all along: I was hardly ever aware of what was going on. It felt like I was walking backward—I wasn't able to see where I was going until I got there. I never knew what was happening until it had finished happening.

Often I was not even aware that I was thinking about the issue. I would be doing something else—washing the dishes, driving to the store—and without the slightest warning a thought concerning quantum mechanics would pop into my head. A pain in the neck? A delight? Yes—yes to both.

Writing was helpful. It always clarifies my thoughts to get something down in plain English. That's why I wrote this book: to clarify my thoughts. Of course I couldn't write the book until I had cleared up my thinking—and I couldn't clear up my thinking without writing the book. So it was a back-and-forth process.

Looking back on it all, I now feel that what I was doing was facing for the first time how utterly strange quantum theory is. A student learning quantum theory must learn a whole new way of doing things. The way involves mathematics that seems to have nothing to do with the subject at hand. An example is those matrices of chapter 6—nowhere within them do you find the slightest image of a spinning object. And the same is true of all the rest of the theory.

And then one day I had an epiphany.

The amazing thing about that epiphany is that it happened in a flash—at a precise instant of time. As a matter of fact, so momentous was that instant that I took note of it. Even now a small sign sits above my desk:

The Epiphany

11 AM, Friday July 10, 2015

Another bright and sunny day

(Big thunderstorm last night)

I don't want to give the wrong impression. I don't want to imply that in order to understand something hard, all you have to do is sit around and wait for inspiration to hit. My epiphany would never have come had I not spent all those years of work stewing things over. The epiphany was just the final step.

Nevertheless, it was a climactic moment. It felt as if I had been wandering around for years through a darkened house, and that I had ultimately found myself in a pitch-black room, a room I had never been in before—and suddenly the lights turned on.

Here is what I saw.

I saw what had been confusing me so thoroughly. It was that I had developed in my mind two completely different spheres of thought. One was the new language of quantum mechanics that I had learned so many years ago. The other was the normal way of thinking that we all employ: the automobile is right *there* and it is going *that way* at *such-and-such* a speed; the golf ball is spinning *so fast* in *this direction*. And what I suddenly realized was that all along I have been thinking in both ways at once. I was moving seamlessly and smoothly from one sphere of thought to the other. And most important of all: this moving from one to the other was unconscious.

If something is unconscious it just might cause you trouble. That, I suddenly saw, was what had been giving me so much grief for so very long.

Indeed, I had been actively preventing myself from realizing how utterly incompatible those two ways of thinking are. This incompatibility is the very essence of this book. It is the essence of Bell's Theorem. What Bell's Theorem proves is that quantum mechanics is not a local hidden-variable theory—and that's just a fancy way to say that it is not a theory of normal reality. And the experiments testing Bell's Theorem—metaphysical experiments—are telling us that the hypothesis of normal reality is untenable. There ain't no such thing.

This "doublethink" had been infecting all my thinking over the years. Indeed, it has infected this very book. In the first chapter I wrote of my youthful amazement that something could be in two places at once. Indeed, I was frustrated that the Great Predictor refused to tell me how this could be. Later on, in chapter 6, I had advanced in my thinking somewhat, and wrote that the problem lay with the language the Great Predictor spoke. It was, I wrote, a strangely impoverished language, and my poor Predictor was simply tongue-tied: his language was incapable of expressing certain things. But now I suddenly realize that the truth is far stranger than that: it is that my very question was misguided. "How can a thing be in two places at once?" I had asked—but buried within that question is an assumption, the assumption that a thing can be in *one* place at once. That is an example of doublethink, of importing into the world of quantum mechanics our normal conception of reality—for the location of an object is a hidden variable, a property of the object ... and the new science of experimental metaphysics has taught us that hidden variables do not exist.

Another example of how I had been unconsciously moving between these two spheres: radioactive decay. In chapter 2 I discussed how one nucleus would decay rapidly while another would decay more slowly. What enables one nucleus to survive for longer than the other? The Great

Predictor refused to say. I was frustrated by this refusal, but now I see that contained within my frustration was another assumption: the assumption that *there is a reason*—some property that distinguishes the short-lived nucleus from the long-lived one. And now I realize that reasons are hidden variables ... and hidden variables do not exist.

I would not be wasting the reader's time on my own personal history had I not felt that it has a wider moral. For in truth I believe that what I have been recounting in this book is not just my own story: I believe it is every scientist's story. Einstein believed utterly in a real physical situation, and he fought for that view to the end of his life. Bell did too: "Everything has definite properties,"[1] he would often say. And perhaps you recall my earlier quote from the quantum physicist E. T. Jaynes, who termed the view that there was no such reality "a violent irrationality" (chapter 13) So I am not ashamed of thinking according to our normal conception of reality. That is how these people thought, and if it was good enough for them it is good enough for me.

As a matter of fact, I believe that the real point goes beyond what I myself think, or what this person or that person thinks. I believe that in truth we cannot help thinking that way. That is the only way we know how to think. It is how our minds work.

And it is how science works—all of science: biology and geology and chemistry and, indeed, every facet of physics other than quantum mechanics. I want to emphasize this. *Never* before have we encountered a situation like the one that experimental metaphysics has forced on us. Relativity, the space program, the genomic revolution, artificial intelligence—none of these have required so great a shift in our thinking. Earlier in my career I worked on neutron stars: monstrously dense, exotic in composition, ferociously magnetic ... but each one of which sits in a perfectly definite place and spins at a perfectly definite rate in a perfectly definite direction. A geologist might be concerned with the motion of magma hundreds of miles beneath her feet—a magma that neither she nor anybody else has ever seen ... but nevertheless a magma that has a perfectly definite temperature and pressure. A biologist might study the evolution of creatures now extinct: creatures that no one has ever seen ... but creatures whose size and shape and mating habits most definitely existed.

No matter how exotic and unfamiliar the objects that scientists study, until now all of them have conformed to our normal conception of reality.

There is only one small problem: the new science of experimental metaphysics has shown that within the microworld this normal conception of reality does not apply.

"Shut up and calculate." That is the way people often refer to the standard approach to quantum mechanics. "Don't waste time thinking about all this stuff" might be a good translation. "Just get going and do the calculations." I used to think the phrase was pejorative. Now I am not so sure. Maybe it is not pejorative. Maybe it is great wisdom.

The astute reader will have noticed that I have not solved the mystery of quantum theory. I have not explained how things can have no properties. I have not explained how nonlocality can bind the universe together. No matter: I am content to rest. At long last, I have achieved what to me is a great victory. I have expressed to myself clearly what the mystery is.

Because in truth I wonder if it *is* a mystery. Perhaps it is just a fact. This is the way the world is.

Do I like this new cosmos that we have stumbled into? Do I dislike it? Is it congenial to my thoughts, or utterly alien to them? Well, I guess I would say that it makes no difference: this is the new world—get used to it.

Listen to the words of Richard Feynman:

> We always have had a great deal of difficulty understanding the world view that quantum mechanics represents. At least I do, because I'm an old enough man that I haven't got to the point that this stuff is obvious to me. Okay, I still get nervous with it. ... You know how it always is, every new idea, it takes a generation or two until it becomes obvious that there's no real problem. It has not yet become obvious to me that there's no real problem. I cannot define the real problem, therefore I suspect there's no real problem, but I'm not sure there's no real problem.[2]

Not long ago I had a dream. In that dream I was on a powerboat far out to sea. The engine was off. No breeze blew: we drifted aimlessly. An immense silence reigned. The sky was gray and vague, the horizon obscured by haze. Nothing was happening. Everything was listless.

At long last I roused myself to wonder where we were. Reaching down into the water I gave a sideways paddle. Slowly the boat spun about—and suddenly there came into view a stone jetty. It was a mere few feet away! While I had been listlessly waiting, the boat had drifted right up against the shore.

The boulders of the jetty were hard and clear, utterly solid and picked out in vivid relief. Looking upward I saw that the haze had lifted, and that the sky was now a crystalline blue. Gazing down into the water I saw that it too was clear and lovely. I could see the bottom. Could I reach it and so give us a push? Leaning over, I found the water just slightly too deep. Or could I reach the jetty, and push off against it? Leaning sideways, I found it just slightly too far away.

Not a problem—we had an engine. I reached for the starter switch.

There was a mirror. I looked at my reflection in it. "Time to get going" I told my reflection.

And I did. And we are.

*The epigraph of this book is a quotation from a lecture by Richard Feynman. Actually, I took that quote out of context. Here is a bit more of it:*

> I think I can safely say that nobody understands quantum mechanics. So do not take the lecture too seriously, feeling that you really have to understand in terms of some model what I am going to describe, but just relax and enjoy it. I am going to tell you what nature behaves like. If you will simply admit that maybe she does behave like this, you will find her a delightful, entrancing thing.[1]

## Appendix 1: The GHZ Theorem

It is a complicated matter even to write down Bell's Theorem—the particular mathematical relation between the various quantities that he proved every hidden-variable theory must obey. And the proof of his theorem is harder still. But some time after Bell's work, an extension of his theorem was discovered—an extension so simple that it is actually possible to describe the theorem, and to give a nontechnical proof.

The authors of this new theorem are Daniel Greenberger, Michael Horne, and Anton Zeilinger—hence their result's name: the GHZ theorem. It is an extension of Bell's work. Recall that Bell had envisaged two particles: in the GHZ analysis there are three. Similarly, the old argument had envisaged two experimenters, Alice and Bob, while the new one has a third—Chris, let's say. In both theorems the experimenters measure the spins of the particles heading toward them and, as with all quantum measurements of spin, there are only two possible results: the spin is found to lie either along the reference direction of the detector, or against it.

In Bell's scenario the particles were in a special entangled state in which Alice and Bob's measurements yielded opposite results if the detectors were parallel. So in this configuration Alice was able to predict the result of Bob's measurement prior to his making it—it would be the opposite of what she had obtained. Recall the central point of the old EPR argument that had so disturbed Bohr: Einstein and his coworkers claimed that this demonstrated that the spin of Bob's particle must have existed all along, contrary to the principles of quantum mechanics.

Similarly, in the Greenberger, Horne, and Zeilinger scenario the particles are also in an entangled state, so the results of measurements can also be predicted and the same conclusion would follow. And just as Bell had found a way to experimentally test the EPR conclusion, so too do Greenberger and colleagues. Thus experimental metaphysics.

The GHZ scenario goes as follows. Alice, Bob, and Chris measure the components of spin of the particles entering their detectors, and each of them writes down on a slip of paper the result. They do so adopting a particular shorthand:

- If the result was that the spin lay *along* the detector reference direction, write down the number one.
- If the spin lay *against* the direction, write down minus one.

Finally, the three experimenters collect their results and multiply them together. The result is itself a single number: let us call it their final combined result. What might this combined result be? It can only be plus or minus one, since each of its three individual components were either plus or minus one.

Greenberger, Horne, and Zeilinger envisage measurements in which the experimenters orient their detectors so that one is horizontal, and the other two vertical. In this particular configuration their entangled state has an important property: the final combined result can only be plus one. It can never be minus one.

Follow now the Einstein, Podolsky, and Rosen argument and see what it makes of this. Imagine with them that Alice moves far off into the distance, so that Bob and Chris make their measurements before she makes hers. Then they can predict the result she will get! For suppose that Bob had obtained, say, spin along the reference direction, and so jotted down +1, while Chris had found spin against it and so written –1. Then, since the product of all three results must be +1, they know that, when Alice makes her measurement, she is sure to obtain –1. So Bob and Chris have determined the spin of Alice's particle: its spin is against her axis.

The same is true for the various other results Bob and Chris might have obtained. So the usual EPR argument leads to the conclusion that the hidden variable corresponding to Alice's measurement exists.

Greenberger, Horne, and Zeilinger realized that they could test this conclusion. They would do so as follows.

Notice that there are three different ways in which Alice, Bob, and Chris can orient their detectors. Two are vertical, and only one of them horizontal—but which is the horizontal one? It might be Alice's. Alternatively, it might be Bob's that is horizontal, or finally Chris's. So there are three possible cases.

In the first case we can write the final combined result as

$$A_{\text{horizontal}}\ B_{\text{vertical}}\ C_{\text{vertical}}$$

where by "A" we mean the number Alice wrote down, "B" the number Bob wrote down, and so too for "C."

In the second case the final result is

$$A_{\text{vertical}}\ B_{\text{horizontal}}\ C_{\text{vertical}}$$

And in the third

$$A_{\text{vertical}}\ B_{\text{vertical}}\ C_{\text{horizontal}}$$

Each of these expressions is the product of three things—the three results that Alice, Bob, and Chris had jotted down on their slips of paper. Recall that these results (plus or minus one) are just shorthand ways to express the fact that the electron's spin lay along or against the direction of their detectors. And recall that the usual EPR argument was that they must have been real properties of the electrons. If this is so—and here is the hidden-variable hypothesis in action—then we can treat these three results as simple numbers, and we can rearrange them in any way we wish. That is what Greenberger and coworkers proceeded to do.

They knew that in their special entangled state each of these products equals plus one. So if they were to multiply them together they would still get plus one:

$$(A_{\text{horizontal}}\ B_{\text{vertical}}\ C_{\text{vertical}})\ (A_{\text{vertical}}\ B_{\text{horizontal}}\ C_{\text{vertical}})\ (A_{\text{vertical}}\ B_{\text{vertical}}\ C_{\text{horizontal}}) = +1$$

They rearranged this:

$$A_{\text{horizontal}}\ B_{\text{horizontal}}\ C_{\text{horizontal}}\ (A_{\text{vertical}})^2\ (B_{\text{vertical}})^2\ (C_{\text{vertical}})^2 = +1$$

And they recalled that no matter what the three experimenters obtained, the numbers they wrote down—A, B, and C—could only be plus or minus one. But the square of minus one is plus one, and so of course is the square of plus one. So they realized that they could simply leave out the terms involving the parentheses—they were just plus one. And they got a most interesting result:

$$A_{\text{horizontal}}\ B_{\text{horizontal}}\ C_{\text{horizontal}} = +1$$

In words: if all three experimenters were to orient their detectors horizontally, their final combined result is guaranteed to be plus one. This is a prediction of the hidden-variables theory, and it can be tested.

There's more. Greenberger, Horne, and Zeilinger realized that quantum mechanics makes the opposite prediction:

$A_{horizontal}\ B_{horizontal}\ C_{horizontal} = -1$

So the final combined result of three horizontal measurements is sure to be minus one. So, just as in Bell's analysis, quantum mechanics disagrees with the postulate of hidden variables. Here too, an experiment to measure this result of this configuration would have metaphysical implications.

That experiment has been done. Its results agree with quantum mechanics, and not with the hidden-variable hypothesis.

## Appendix 2: Further Reading

If you find your interest piqued by this little book, you may want to explore the subject further. Here I list some good readings. All include a discussion of Bell's Theorem, although in some cases only tangentially.

### Non-Technical Works

The following are specifically aimed at a general audience. I can testify from personal experience that are all very well written and a pleasure to read

Becker, Adam. *What Is Real? The Unfinished Quest for the Meaning of Quantum Physics.* New York: Basic Books, 2018.

> As the title makes clear, this book is primarily concerned with attempts to understand what quantum theory is telling us about the world. It contains many personal anecdotes, and also much about the three "alternative" approaches that I mentioned in chapter 2

Bernstein, Jeremy. "John Stewart Bell: Quantum engineer," in Jeremy Bernstein, *Quantum Profiles.* Princeton: Princeton University Press, 1991, pp. 3–91.

> Far broader in scope than most of the other readings I have listed here, Bernstein surveys the entire field of quantum theory and its many mysteries.

Gilder, Louisa. *The Age of Entanglement: When Quantum Physics was Reborn.* New York: Alfred A. Knopf, 2008.

> With a novelist's skill, Gilder humanizes and dramatizes the story of the discovery of entanglement. Much of the material in my chapter 11 is taken from this book.

Herbert, Nick. *Quantum Reality: Beyond the New Physics.* New York: Anchor Press / Doubleday, 1985.

> A broad survey of quantum theory's mysteries.

Kaiser, David. *How the Hippies Saved Physics: Science, Counterculture and the Quantum Revival.* New York: W. W. Norton, 2011.

Kaiser is a physicist and a historian of science: the title of his book tells it all. Much of the material in my chapters 4 and 12 is taken from this book.

Kaiser, David. "Quantum theory by starlight." *New Yorker* (online), February 7, 2017, available at https://www.newyorker.com/tech/elements/quantum-theory-by-starlight

Kaiser describes here the experiment that closed the freedom of choice loophole (in which he participated).

Kaiser, David. "Free will, video games, and the most profound quantum mystery." *New Yorker* (online), May 9, 2018, available at https://www.newyorker.com/science/elements/free-will-video-games-and-the-most-profound-quantum-mystery

Kaiser describes here the experiment that used a video game to elicit random choices from people worldwide to be used in Bell-test experiments.

Zeilinger, Anton. *Dance of the Photons: From Einstein to Quantum Teleportation.* New York: Farrar, Straus and Giroux, 2010.

Zeilinger, a physicist at the forefront of contemporary experimental research on entanglement and Bell's Theorem, provides an insider's view of the field. Much of the material in my chapter 15 is taken from this book.

## Semitechnical Books

You probably would need a certain amount of technical background to fully engage with the following books, but they are not solely for experts.

Freire, Olival. *Quantum Dissidents: Rebuilding the Foundations of Quantum Mechanics.* Berlin: Springer, 2015.

Friere is a historian of science. This book is particularly revealing on the "stigma" attached to thinking about the foundations of quantum mechanics.

Lewis, Peter J. *Quantum Ontology: A Guide to the Metaphysics of Quantum Mechanics.* Oxford: Oxford University Press, 2016.

Lewis is a philosopher: the title tells it all.

Whitaker, Andrew. *John Stewart Bell and Twentieth-Century Physics.* Oxford: Oxford University Press, 2016.

A detailed look at Bell's life and work, written by a physicist.

## Other Forms of Bell's Theorem

As I commented in appendix 1 on the GHZ theorem, since Bell's discovery other forms of his theorem have been found. And as I commented in chapter 10, Bell's Theorem is not really about physics at all. It is pure logic: a matter of analyzing all the ways a random variable can be distributed.

And in particular, it is not even solely about spin. Two different works have described a form of Bell's Theorem using a "spin-free" approach.

d'Espagnat, Bernard. "The quantum theory and reality." *Scientific American* (November 1, 1979), pp. 158–181.

The above article is written for a general audience. On the other hand, you may find yourself positively scared when you first pick up the following book, since it is intended for an audience of physics students. But do not give up! The specific section listed here requires no special training at all: once again, it is all a matter of pure logic.

Greenstein, George, and Arthur G. Zajonc. *The Quantum Challenge: Modern Research on Quantum Mechanics.* 2nd edition. Burlington, MA: Jones and Bartlett, 2006, pp. 142–148. This describes a method developed by N. David Mermin.

In general, I would guess that if you found yourself comfortable with chapter 9's treatment of a hidden-variable theory, or the appendix's treatment of the GHZ Theorem, you will find yourself equally comfortable with either of the above two references.

But I cannot help but point you toward the article in which Mermin first introduced his new proof. I mentioned this article in chapter 9: it was immensely helpful to me in my efforts to understand Bell's Theorem. Even though it is written for an audience of physicists, a great part of it consists of an utterly delightful history of the early arguments among the founders of quantum theory: you really should take a look at it!

Mermin, N. David. "Is the Moon there when nobody looks?" *Physics Today* 38 (April 1985): 38–47.

# Notes

## Foreword

1. Erwin Schrödinger, "Discussion of probability relations between separated systems," *Mathematical Proceedings of the Cambridge Philosophical Society* 31 (1935): 555–563, on 555.

2. Schrödinger, 555; Albert Einstein, Boris Podolsky, and Nathan Rosen, "Can quantum-mechanical description of physical reality be considered complete?," *Physical Review* 47 (1935): 777–780.

3. Einstein, Podolsky, and Rosen, "Can quantum-mechanical description of physical reality be considered complete?," 777.

4. Niels Bohr, "Can quantum-mechanical description of physical reality be considered complete?," *Physical Review* 48 (1935): 696–702. Abraham Pais recounts the story of Einstein asking him about the moon in Pais, *"Subtle Is the Lord ...": The Science and the Life of Albert Einstein* (New York: Oxford University Press, 1982), pp. 5–6. On the Einstein–Bohr debate, see especially Don Howard, "Einstein on locality and separability," *Studies in History and Philosophy of Science* 16 (1985): 171–201; Don Howard, "'Nicht sein kann was nicht sein darf,' or the prehistory of EPR, 1909–1935: Einstein's early worries about the quantum mechanics of composite systems," in *Sixty-Two Years of Uncertainty: Historical, Philosophical, and Physical Inquiries into the Foundations of Quantum Mechanics*, ed. Arthur I. Miller (New York: Plenum, 1990), pp. 61–112; Arthur Fine, *The Shaky Game: Einstein, Realism, and the Quantum Theory* (Chicago: University of Chicago Press, 1986); and Mara Beller, *Quantum Dialogue: The Making of a Revolution* (Chicago: University of Chicago Press, 1999).

5. See esp. Jeremy Bernstein, "John Stewart Bell: Quantum engineer," in Bernstein, *Quantum Profiles* (Princeton: Princeton University Press, 1991), 3–91; Louisa Gilder, *The Age of Entanglement: When Quantum Physics was Reborn* (New York: Knopf, 2008); David Kaiser, *How the Hippies Saved Physics: Science, Counterculture, and the Quantum Revival* (New York: W. W. Norton, 2011); Olival Freire, Jr., *The Quantum Dissidents: Rebuilding the Foundations of Quantum Mechanics, 1950–1990* (New York: Springer,

2014); and Andrew Whitaker, *John Stewart Bell and Twentieth-Century Physics* (New York: Oxford University Press, 2016).

6. John S. Bell, "On the Einstein Podolsky Rosen paradox," *Physics* 1 (1964): 195–200.

7. See esp. John Clauser, Michael Horne, Abner Shimony, and Richard Holt, "Proposed experiment to test local hidden-variable theories," *Physical Review Letters* 23 (1969): 880–884; Stuart Freedman and John Clauser, "Experimental test of local hidden-variable theories," *Physical Review Letters* 28 (1972): 938–941. On the early experimental efforts to test Bell's inequality, see also Gilder, *Age of Entanglement*, chapters 30–31; Kaiser, *How the Hippies Saved Physics*, chapters 3 and 8; and Freire, *Quantum Dissidents*, chapters 7–8.

8. John Trimmer, "The present situation in quantum mechanics: A translation of Schrödinger's 'Cat Paradox' paper," *Proceedings of the American Philosophical Society* 124 (1980): 323–338. Schrödinger's paper was originally published in three installments as "Die gegenwärtige Situation in der Quantenmechanik," *Naturwissenschaften* 23 (1935): 807–812, 823–828, and 844–849. For recent discussion of loopholes in experimental tests of Bell's inequality, see J.-A. Larsson, "Loopholes in Bell inequality tests of local realism," *Journal of Physics A* 47 (2014): 424003; and N. Brunner, D. Cavalcanti, S. Pironio, V. Scarani, and S. Wehner, "Bell nonlocality," *Reviews of Modern Physics* 86 (2014): 419–478.

9. The first loophole to be addressed experimentally concerns "locality," that is, whether information from one side of the apparatus could have affected the other side during the conduct of a given experiment. See Alain Aspect, Jean Dalibard, and Gérard Roger, "Experimental test of Bell's inequalities using time-varying analyzers," *Physical Review Letters* 49 (1982): 1804–1807; Gregor Weihs, Thomas Jennewein, Christoph Simon, Harald Weinfurter, and Anton Zeilinger, "Violation of Bell's inequality under strict Einstein locality conditions," *Physical Review Letters* 81 (1998): 5039–5043.

10. B. Hensen et al., "Loophole-free Bell inequality violation using electron spins separated by 1.3 kilometers," *Nature* 526 (2015): 682–686; M. Giustina et al., "Significant-loophole-free test of Bell's theorem with entangled photons," *Physical Review Letters* 115 (2015): 250401; L. K. Shalm et al., "Strong loophole-free test of local realism," *Physical Review Letters* 115 (2015): 250402; W. Rosenfeld et al., "Event-ready Bell test using entangled atoms simultaneously closing detection and locality loopholes," *Physical Review Letters* 119 (2017): 010402; and M.-H. Li et al., "Test of local realism into the past without detection and locality loopholes," *Physical Review Letters* 121 (2018): 080404.

11. Jason Gallicchio, Andrew Friedman, and David Kaiser, "Testing Bell's inequality with cosmic photons: Closing the setting-independence loophole," *Physical Review Letters* 112 (2014): 110405; Johannes Handsteiner et al., "Cosmic Bell test: Measurement settings from Milky Way stars," *Physical Review Letters* 118 (2017): 060401; Dominik Rauch et al., "Cosmic Bell test using random measurement settings from high-redshift

quasars," *Physical Review Letters* 121 (2018): 080403. See also C. Abellán et al. (The Big Bell Test collaboration), "Challenging local realism with human choices," *Nature* 557 (2018): 212–216.

12. Jason Palmer, "Subatomic opportunities: Quantum leaps," *Economist* (March 9, 2017).

13. Kurt Jacobs and Howard Wiseman, "An entangled web of crime: Bell's Theorem as a short story," *American Journal of Physics* 73 (2005): 932–937; Paul Kwiat and Lucien Hardy, "The mystery of the quantum cakes," *American Journal of Physics* 68 (2000): 33–36; Seth Lloyd, *Programming the Universe: A Quantum Computer Scientist takes on the Cosmos* (New York: Knopf, 2006), 120–121; see also Kaiser, *How the Hippies Saved Physics*, pp. 37–38.

## Chapter 4

1. From a taped discussion with R. Jost on December 2, 1961; quoted in Abraham Pais, *Niels Bohr's Times* (New York: Oxford University Press, 1991), p. 318.

2. L. Rosenfeld in *Proceedings of the 14th Solvay Conference* (New York: Interscience, 1968), p. 232.

3. Letter from Einstein to Schrodinger May 31, 1928, reprinted in "Letters on Wave Mechanics" ed. M. Klein (New York: Philosophical Library, 1967).

4. Letter to Paul Ehrenfest, quoted in Abraham Pais, *Subtle Is the Lord* (New York: Oxford University Press, 1982), pp. 416–417.

5. David Kaiser, *How the Hippies Saved Physics* (New York: W. W. Norton, 2011), p. 164.

6. Kaiser, p. 143.

7. John Clauser, "Early history of Bell's Theorem," in *Quantum [Un]speakables: From Bell to Quantum Information*, ed. R. A. Bertlmann and A. Zeilinger (Berlin: Springer-Verlag, 2002), p. 72.

## Chapter 7

1. Einstein, Podolsky, and Rosen, "Can quantum-mechanical description of physical reality be considered complete?," p. 777.

2. Leon Rosenfeld, in S. Rozental, ed., *Niels Bohr, His Life and Work as Seen by His Friends and Colleagues*, (Amsterdam: North Holland, 1967), pp. 128–129.

3. Andrew Whitaker in Bertlmann and Zeilinger, *Quantum [Un]speakables*, p. 15.

4. In Graham Farmelo, "Random acts of science," *New York Times Sunday Book Review*, June 11, 2010.

5. In Douglas Huff and Omer Prewett, eds., *The Nature of the Physical Universe: 1976 Nobel Conference* (New York: John Wiley & Sons, 1979), p. 29.

### Chapter 10

1. Reinhold A. Bertlmann in *Quantum [Un]speakables*, ed. R. A. Bertlmann and A. Zeilinger (Berlin: Springer-Verlag, 2002), p. 29.

2. Abner Shimony in Bertlmann and Zeilinger, *Quantum [Un]speakables*, p. 55.

3. M. Bell, K. Gottfried, and M. Veltman, eds., *John S. Bell on the Foundations of Quantum Mechanics* (World Scientific, 2001), p. 216.

4. J. S. Bell, *Speakable and Unspeakable in Quantum Mechanics* (Cambridge: Cambridge University Press, 1987), p. 28.

5. Bell, p. 2.

6. Bertlmann and Zeilinger, *Quantum [Un]speakables*, p. 295.

7. George Greenstein and Arthur G. Zajonc, *The Quantum Challenge: Modern Research on the Foundations of Quantum Mechanics*, 2nd ed. (Burlington, MA: Jones and Bartlett, 2006), pp. 142–148.

### Chapter 11

1. David Kaiser, "How the hippies saved physics" (New York: W. W. Norton, 2011), p. xiv.

### Chapter 12

1. Louisa Gilder, *The Age of Entanglement* (New York: Alfred A. Knopf, 2008), p. 253.

2. Interview of John Clauser by Joan Bromberg on May 20, 2002, Niels Bohr Library & Archives, American Institute of Physics, College Park, MD, www.aip.org/history-programs/niels-bohr-library/oral-histories/25096 (hereafter "AIP Interview").

3. AIP Interview.

4. AIP Interview.

5. Gilder, *The Age of Entanglement*, p. 261.

6. AIP Interview.

7. AIP Interview.

8. Clauser, "Early history of Bell's Theorem," p. 80.

9. AIP Interview.

10. AIP Interview.

11. Gilder, *The Age of Entanglement*, pp. 267–268.

12. Gilder, p. 267.

13. AIP Interview.

14. AIP Interview.

15. Clauser, "Early history of Bell's Theorem," p. 62.

16. Clauser, p. 72.

17. David Kaiser "Quantum theory by starlight," *New Yorker* (online), February 7, 2017.

## Chapter 13

1. AIP Interview.

2. E. T. Jaynes, "Quantum Beats," in *Foundations of Radiation Theory and Quantum Electrodynamics*, ed. A. O. Barut (New York: Plenum Press, 1980).

## Chapter 16

1. Reinhold A. Bertlmann, "Magic Moments with John Bell," *Physics Today* 68, no. 7 (July 2015): 40.

2. Richard Feynman, "Simulating physics with computers, *International Journal of Theoretical Physics* 21, nos. 6/7 (1982): 467–488.

## Epigraph

1. Richard Feynman, *The Character of Physical Law* (Cambridge, MA: MIT Press, 1965), p. 129.

# Index

Numbers in italics indicate images.


Abellan, Carlos, 85, *85,* 85–90
Abstract thinking, 34, 35
Aspect, Alain, 79–82, *80,* 85, 93
Atomic bomb, 66, 104
"Aunt Martha's Coffin," 105

Bank transfers, 102–106
  Bell, John, 2–7, *58,* 101, 117
  background of, 57–58
  combination of mathematical quantities by, 61–62
  GHZ theorem and, 121–122
  Greenstein's meeting with, 59–60
  indifference over work of, 65–66
  passion and commitment of, 60–61
  pioneering work of, 61
  work of, 55–62
Bell, Mary, *58*
Bell's Theorem, 2–3, 26, 81
  Alain Aspect and, 79–82, 85
  Anton Zeilinger and, 82–85
  experiments based on, 63–64
  George Greenstein's early understanding of, 45–46, 113–117
  as hidden variable theory, 46–53
  hidden variable theory and (*See* Hidden variable theory)
  John Bell's work and, 55–62
  John Clauser and, 71–79
  locality assumption in, 94–96
  nature violating, 93
  nonlocality and, 97–100
  other forms of, 127
  randomness and, 85–90
  scientists' desire to close loopholes in, 90–91
Bennett, Charles, 104–105, 106
Bohm, David, 98
Bohr, Niels, 22–26, *25,* 37, 40, 66, 101, 121
Brassard, Gilles, 104–105, 106

Can Quantum-Mechanical Description of Physical Reality Be Considered Complete?," 37
Clauser, John, 71–79, *72*
  desire to discredit quantum theory, 93
CNN, 98
Cold War, the, 66–67, 101
Computers, quantum, 109–111

Defense Intelligence Agency, 110

Einstein, Albert, 6, 17–19, *18, 25,* 30, 35, 101, 117, 121, 122
  attack on Heisenberg's uncertainty principle, 21–27
  EPR paradox and, 37–40

Einstein, Albert (cont.)
  theory of relativity of, 99
  thought experiments by, 22, 24, 26
Electromagnetism, 17
Electrons, nature of, 22
Electron spin, 29–31
EPR paradox and, 37–38
  hidden variable theory, 46–53
  locality assumption, 94
  matrix of, 33–34
  transfering information, 102–105
Encryption, 104–105
"Enigma" machine, 104
  EPR paradox, 37–40
  hidden variable theory, 46–53
  nonlocality and, 97
Esalen Institute, 77
ESP phenomena, 110
Experimental metaphysics, 5–6
  encryption in, 104–105
  new age of, 101–102
  quantum teleportation, 106–111

Feynman, Richard, 118, 120
Freedom of choice experiment, 82–85, 105–106
Fundamental Fysiks Group, 76

Gell-Mann, Murray, 40
GHZ theorem, 121–124
Google, 101
Gravitation, theory of, 17
Great Predictor, 1, 3–5, 21, 35
  EPR paradox and, 37–40, 46–53
  Greenstein's first encounter with, 45, 113
  hidden variables and, 41–42
  silence of, 11–12, 14, 116
  *See also* Quantum mechanics
Greenberger, Daniel, 121
Greenstein, George
  early understanding of Bell's Theorem, 45–46, 113–117
  fascination with Bell's Theorem, 2–3, 15, 115–116
  first encounter with the Great Predictor, 45, 113
  meeting with John Bell, 59–60
  on scientific view of reality, 117–119
Half-life, 41–42
Heisenberg, Werner, 21, 22
Hidden variable theory, 41–42, 46–53, 71, 116
  Alain Aspect and, 79–82, 85
  Anton Zeilinger and, 82–85
  freedom of choice and, 81–82
  John Bell's work and, 55–62
  John Clauser and, 71–79
  locality assumption in, 94
  radioactive decay and, 41–42, 116–117
  randomness and, 85–90
  *things* in, 94–96
Horne, Michael, 121
Hubble Space Telescope, 7

Intel Corporation, 109

Jaynes, E. T., 96, 117

Kaiser, David, 66

Language of quantum mechanics, 33–35
Large Hadron Collider, 7
Locality assumption, 94–96

*Mathematical Foundations of Quantum Mechanics*, 65
Matrix, 33–34
McCarthyism, 78
Metaphysics. *See* Experimental metaphysics
Micus Satellite, *107*, 109
Mitchell, Morgan, 85, *85*, 85–90
Moore, Gordon, 109
Moore's Law, 109–110

Newton, Isaac, 17
Nonlocality, 97–100, 101, 108

Pan, Jian-Wei, 106, *106, 107*
*Physical Review*, 27
Podolsky, Boris, 37, 39, 122

Quantum computers, 109–111
Quantum machines
information transfer with, 102–105
philosophical differences and, 101–102
quantum computers, 109–111
quantum teleportation and, 106–109
research milestones in, 105–106
Quantum mechanics
as abbreviated language, 33–35
in the Cold War, 66–67
electron spin and (*See* Electron spin)
freedom of choice experiment, 82–85, 105–106
GHZ theorem, 121–124
as Great Predictor, 1, 3–5, 113
Greenstein's fascination with, 2–3, 15, 115–116
as half a theory, 17–19, 27, 34
hidden variables theory, 46–53, 71
John Bell and, 2–7
John Clauser and, 71–79
locality assumption, 94–96
measurements creating a property of the microworld, 96
Moore's Law and d, 109–110
new experimental metaphysics and, 101–102
nonlocality in, 97–100, 101, 108
non-technical works on, 125–126
physicists' arguments about, 13
popular books on, 26–27
questions not answered by, 12–13
scientists' desire to close loopholes in, 90–91
semitechnical books on, 126
significance of, 1
transfering information, 102–105
uncertainty theory in, 21–27
von Neumann's work on, 64–65
*See also* Great Predictor
Quantum teleportation, 106–111

Radioactive decay, 41–42, 116–117
Randomness, 85–90
Reality, scientific concepts of, 117–119
Rosen, Nathan, 37, 39

Solvay, Ernest, 22
Solvay Conference, 22
Spin, electron. *See* Electron spin

Teleportation, quantum, 106–111
Theory of relativity, 99
*Things*, 94–96
Thought experiments, 22, 24, 26

Uncertainty principle, 21

Variables, hidden, 41–42
Von Neumann, John, 64–65, 101

World War II, 66, 104

Zajonc, Arthur, 59
Zeilinger, Anton, 82–85, *83*, 105–106, 108–109, 121, 122

Printed in the United States
by Baker & Taylor Publisher Services